AF820657

MAC-10 COOKBOOK

by: Randal Stepan
Nolan Wilson
Gary Reisewitz

MAC-10 Cookbook

by

Randal Stepan
Nolan Wilson
Gary Reisewitz

215 S. Washington
El Dorado, AR 71731-1751
501-862-2077

ISBN 0-87947-310-X
12 11 10 9 8 7 6
Printed in U. S. A.

Desert Publication is a division of
The DELTA GROUP, Ltd.
Direct all inquiries & orders to the above address.

FOREWORD

Much has been written about the "MAC," but until this "Cookbook" was inspired there was no source that covered (nearly) the entire spectrum of all the MAC's produced.

We have covered all that was available to us. While some MAC "Spin-offs" were not included, the most common models are presented in this book. We even included an old catalog of the many accessories that were once available. We hope you enjoy this book as much as we did putting it together.

TABLE OF CONTENTS

Chapter I
FROM THE BEGINNING
by Nolan Wilson

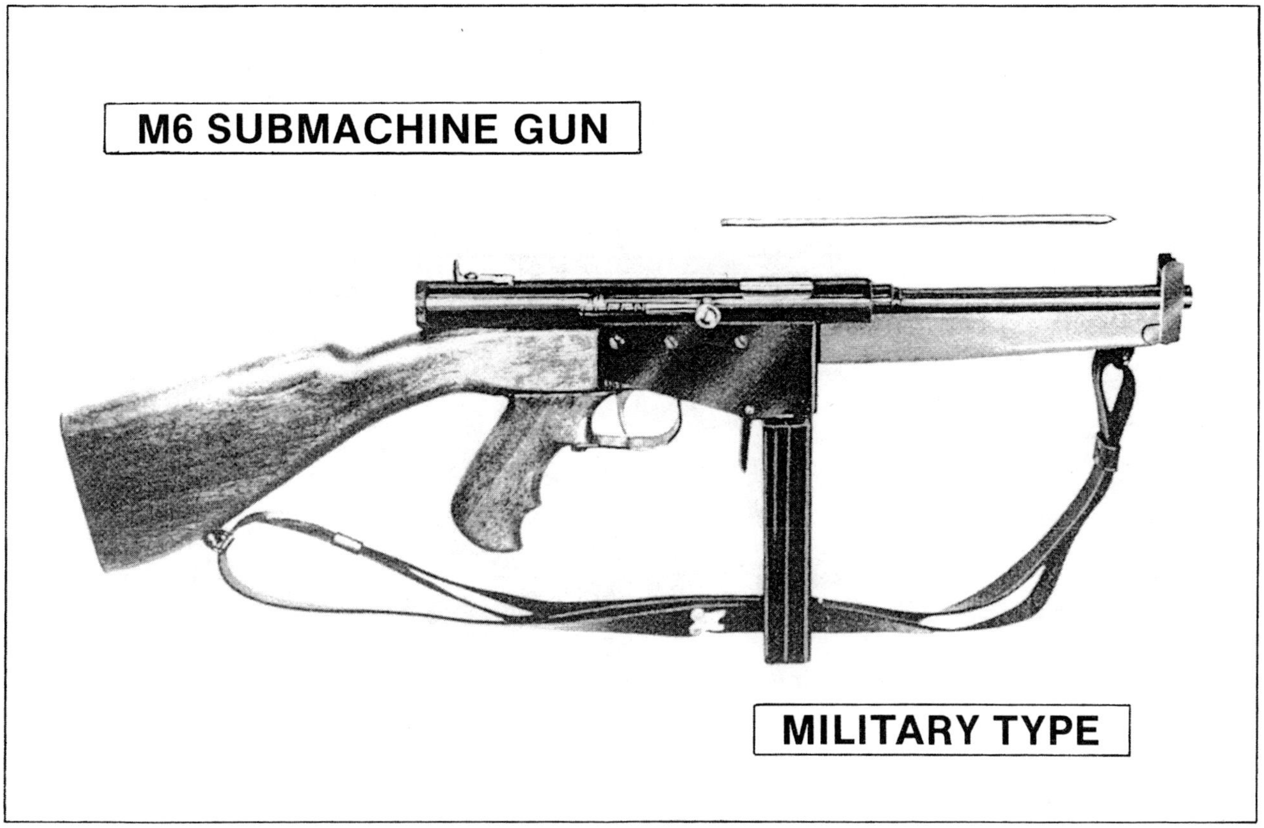

Most people, friend or foe, think the "MAC" to be a weapon of recent design. However, its family tree traces back further than the birth date of many of its fans. One only has to briefly study the work of Gordon Ingram to learn that the Ingram first sprang to life in 1946 in the prototype form of the Model 6. This Model bore little resemblance to the "MAC" of today. In fact it looked more like a child from the marriage of Mr. Thompson to Miss Sten! There was only one prototype of it ever built by the Lightning Arms Company.

Some improvements were made to the first model resulting in the Model 6 being made in three versions: Police, Military and Guard Models. While a few guns were produced in 9mm and .38 Super, the majority (over 15,000) of the Model 6 were produced in caliber .45 ACP by Police Ordnance Company of El Monte, California. This company closed its doors in the early 50's, after which Ingram developed the Model 7 that fired from a closed bolt with both automatic and semiautomatic capabilities. Sales thrust focused on its accurate semiautomatic fire.

The next Ingram, the Model 8, was an improved version of the previous Model 6's, with the Military version featuring an ejection port cover. A Model 9 was the designation of a Model 8, with a collapsible stock added. Although, no production was forthcoming.

With the application of a new design approach, incorporating a magazine well within the pistol grip, the Model 10 was born in the early 70's. Conventionality of both war and weapons was changing rapidly, with clandestine and unconven-

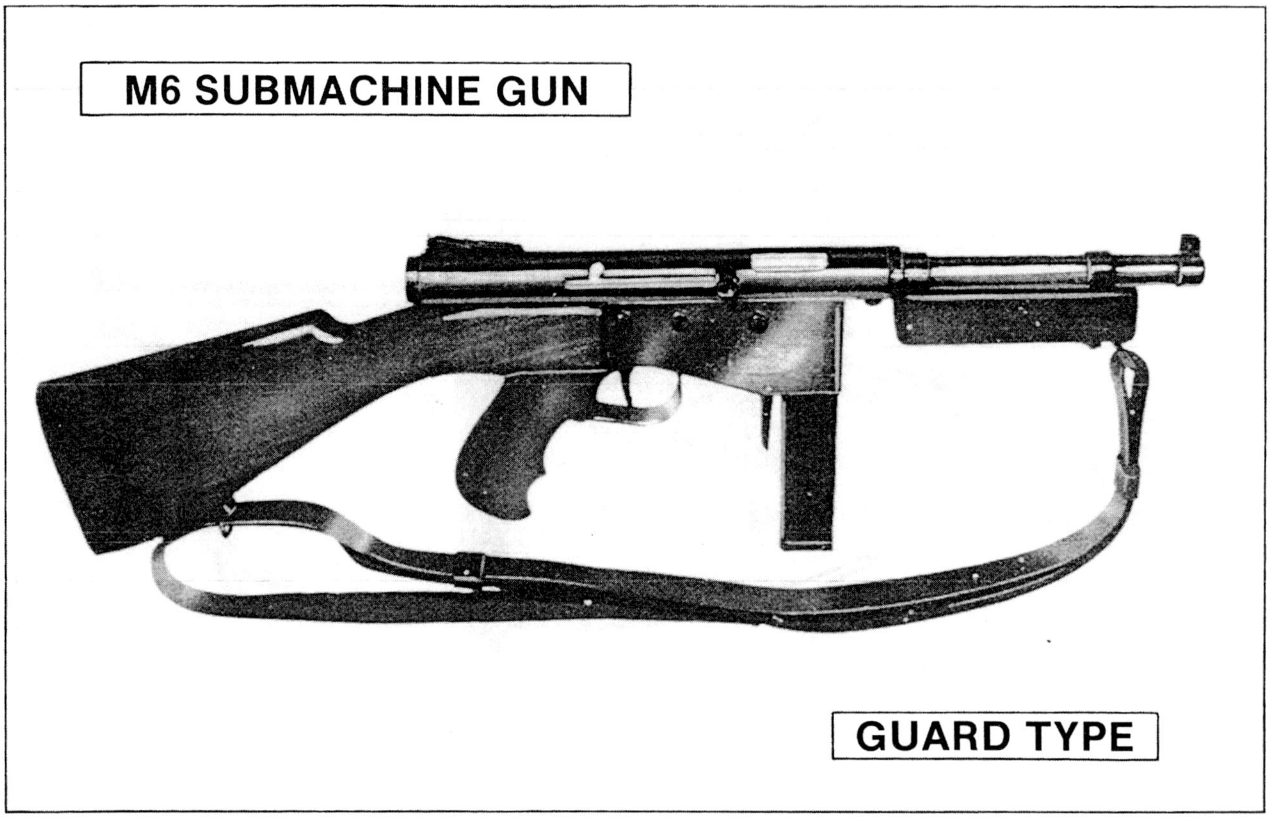

tional warfare demanding more radical approaches to weapons design. Conventional submachine guns, such as the Thompson, gave way to smaller and cheaper to manufacture weapons. Expensive machining, exotic wood and precision fit had little appeal to Third World countries whose technologies were maxed out on sheet metal parts and hand-drilled holes. The new, compact, easily-manufactured Model 10, in both 9mm and .45 ACP, began to find a large market awaiting it, both in the USA and many smaller countries.

While Gordon Ingram was developing his famous guns in California, another weapons genius, the late Mitchell L. WerBell III, was active in a weapons-related business in Georgia. His company, Sionic (Studies in Operational Negation of Insurgency) was busy developing and producing a radically new, suppressor design. Invented by WerBell, the Sionic suppressor was finding military applications in Vietnam with Special Forces and SEALS.

The Model 10 and the Sionic suppressor were a natural for each other. The M10, firing the subsonic .45 ACP, could be quieted to where the bolt rattle made more noise than the round fired. The additional length of the Sionic suppressor, when attached to the M10, was still quite compact and even aided the control of the weapon by giving the left hand a place to hold. It was only natural that WerBell was soon demonstrating the silenced Model 10 to applicable forces in Vietnam. It was, also, a natural that Gordon Ingram was soon moving to Georgia to work with WerBell.

At sometime during the above events,

the Model 11 was born. With clandestine warfare an established science and the acceptance of the Model 10, because of its compactness and concealability, the Model 11 was an instant success. Firing the .380 ACP, it was similar in size to the Colt 1911 pistol. Such firepower from such a small package! A perfect weapon for any "War in the Shadows."

During the early 70's, under the company name of Military Armament Corporation, "MAC's" were in full production at WerBell's Powder Springs, Georgia, facility. It was during this time that many interesting adaptations and attachments were developed, including a grenade launcher. Probably the most noteworthy item was the Operational Briefcase for the M11. It was simply a standard briefcase, with a custom interior fitted to accept a Model 11 submachine gun with a Sionic suppressor attached. The weapon was fired by manipulating a lever from the bottom of the case. The most novel part of this case was where the bullets exited. This opening, which would have exposed the end of the suppressor, was outlined by a small metal frame that allowed a business card to be inserted. Nice touch! It came as no large surprise when the BATF declared the case to be an assassination device requiring registration.

During the mid 70's, many internal problems began to plague the Military Armament Corporation to the point that both Ingram and WerBell left the company. Until this time, the weapons produced were known as "Ingram Submachine Guns." After the split-up, the company changed the weapon's name to "MAC Submachine Gun." All efforts of survival seemed to fail, resulting in the

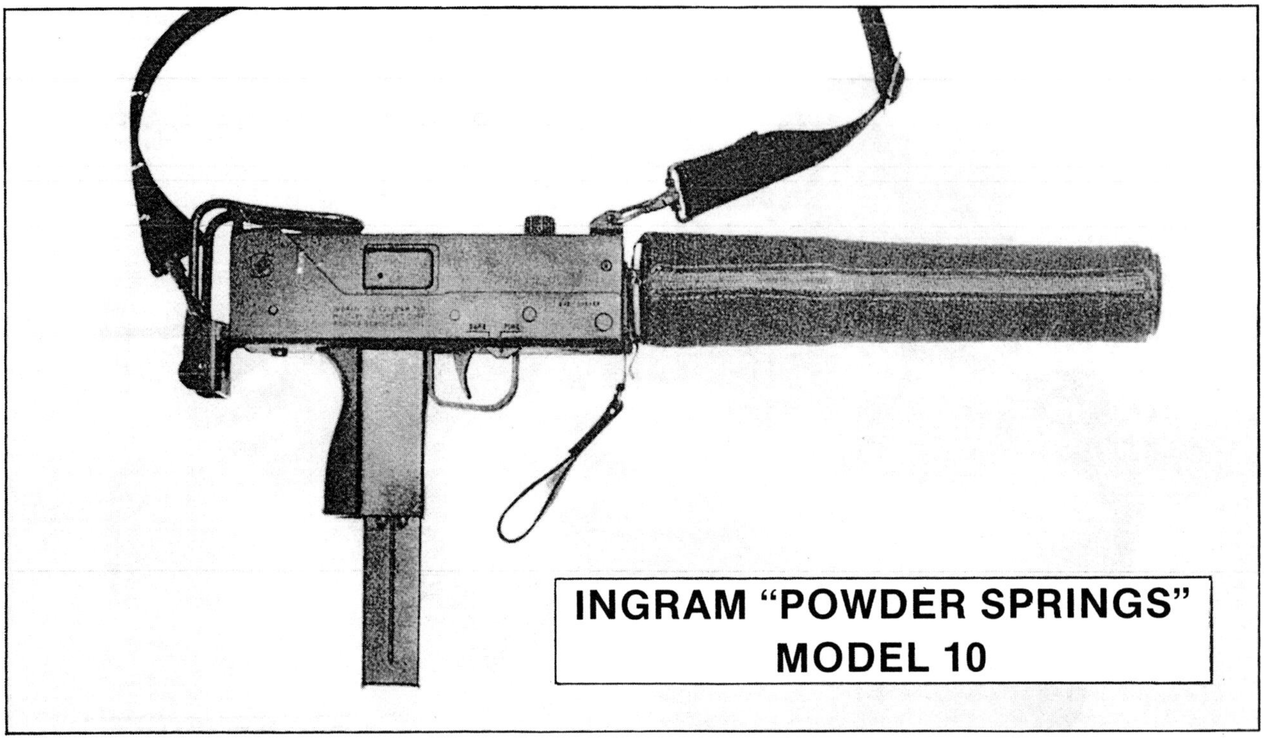

INGRAM "POWDER SPRINGS" MODEL 10

final bankruptcy of Military Armament Corporation in 1975.

A Phoenix Bird soon arose in the name of RPB Industries for the continuation of manufacturing both the MAC guns and Sionic suppressors. About the same time, Wayne Daniel, also of Georgia, formed another arms company by the name of Precision Armaments Company. RPB continued to falter until 1978, at which time Wayne Daniel bought RPB's majority stock and, again, the *now famous* MAC guns continued on.

RPB began to focus heavily on the ready civilian market that had evolved from the Vietnam era. This market had its senses continually-tantalized by Hollywood's flaunting of the MAC-10 in such movies as. John Wayne's *McQ,* Robert Redford's *Three Days of the Condor,* plus many more—including TV spook and spy runs. By making the MAC available in semiauto at an attractive price, the "Big MAC Attack" was under way!

This was to be a "good news/bad news" situation. These semi's fired from an open bolt by the addition of a disconnector mechanism, part of which was permanently attached to the receiver. While this arrangement apparently met with BATF approval at first, they were soon focusing attention on the fact that the mechanically clever were easily-converting these to fully automatic capability. At the time such conversions could legally be done if state and local laws allowed, with an approved BATF Form 1. While the majority of the *off the books* conversions were done by a few of the "good old boys" for *come the day,* a few (if not more than a few) found their way into the criminal mainstream, resulting in the BATF ruling that all open bolt firing MAC's, manufactured after June 1982, were considered machine guns. Needless to say, the BATF began to pay a considerable amount of attention to RPB.

Chapter II
MAC MODELS 10 & 11

A. TECHNICAL SPECIFICATIONS

	Model 10 .45 ACP		Model 10 9MM Para.		Model 11 .380 ACP	
GUN DIMENSIONS WITHOUT SUPPRESSOR	Inches	MM	Inches	MM	Inches	MM
Length without stock	10.50	267	10.50	267	8.75	222
Length stock telescoped	11.60	295	11.60	295	9.81	249
Length stock extended	21.57	548	21.57	548	18.11	460
Barrel Length	5.75	146	5.75	146	5.06	129
Maximum width	1.96	50	1.96	50	1.36	35
GUN DIMENSIONS WITH SUPPRESSOR ATTACHED						
Length without stock	20.35	517	20.35	517	16.25	413
Length stock telescoped	21.45	545	21.45	545	17.31	440
Length stock extended	31.42	798	31.42	798	25.61	650
Maximum width	2.13	54	2.13	54	1.75	44
SUPPRESSOR DIMENSIONS						
Length	11.44	291	11.44	291	8.81	224
Diameter	2.13	54	2.13	54	1.75	44
WEIGHT AND CAPACITY	Lbs.	Kg.	Lbs.	Kg.	Lbs.	Kg.
Gun without magazine	6.25	2.81	6.25	2.81	3.50	1.57
Suppressor	1.20	0.54	1.20	0.54	1.00	0.45
Loaded 16 round magazine	—	—	—	—	0.62	0.27
Loaded 30 round magazine	2.15	0.97	—	—	—	—
Loaded 32 round magazine	—	—	1.37	0.61	1.12	0.54

TYPE OF FIRE..Semi-automatic or full automatic

CYCLIC RATE OF FIRE, M10/45 ACP........................Minimum 950 rounds per minute

M10/9MM Para........................Minimum 950 rounds per minute

M11 .380 ACP ....Minimum 1200 rounds per minute

FRONT SIGHT ....Protected post

REAR SIGHT, M10 ....Fixed aperture for 100 meters

M11 ....Fixed aperture for 50 meters

SAFETIES........................Manually operated safeties for locking bolt in open or closed position

STOCK ASSEMBLY

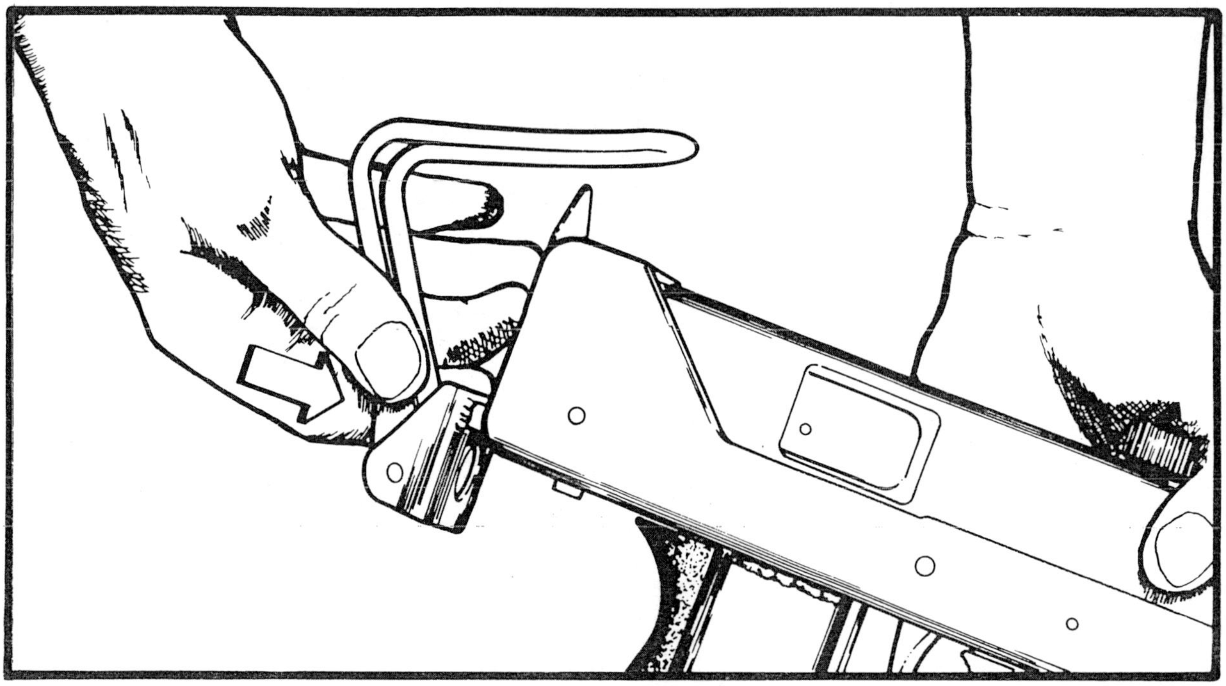

Hold weapon in left hand, with right hand press inwards on right side of wire form butt near the pivot to allow wire form stock to rotate clockwise to normal firing position.

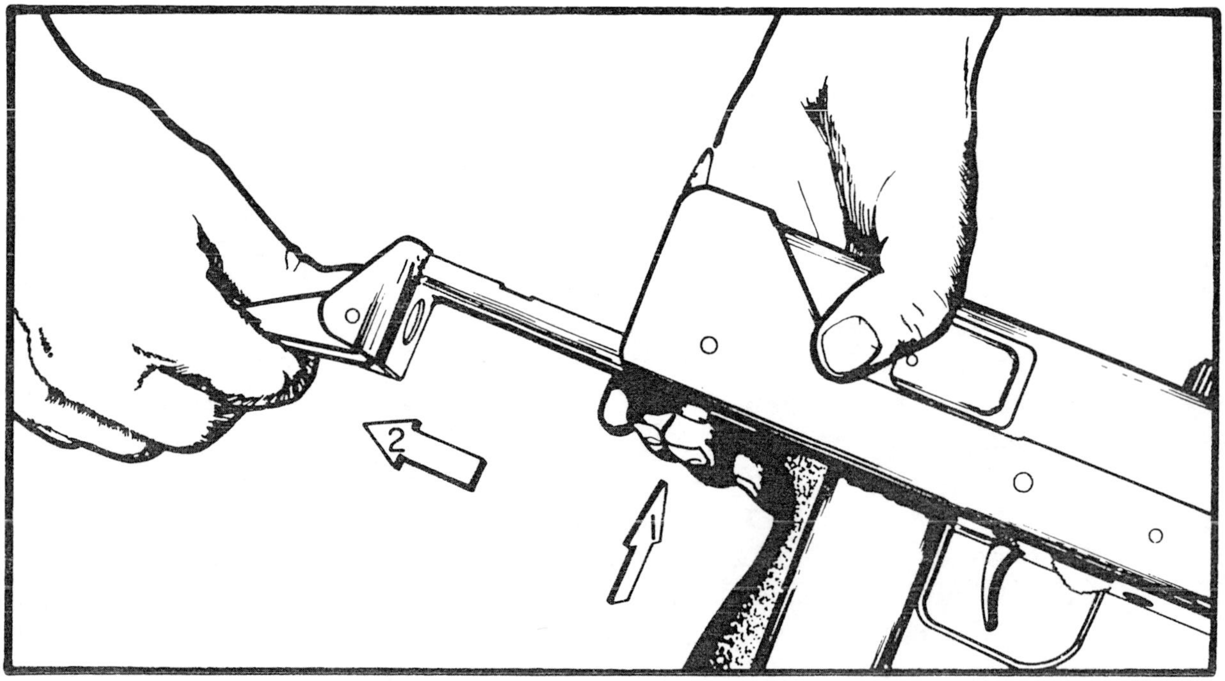

Still holding weapon in left hand depress stock latch button and extend stock assembly using right hand.

LOADING AND COCKING THE WEAPON

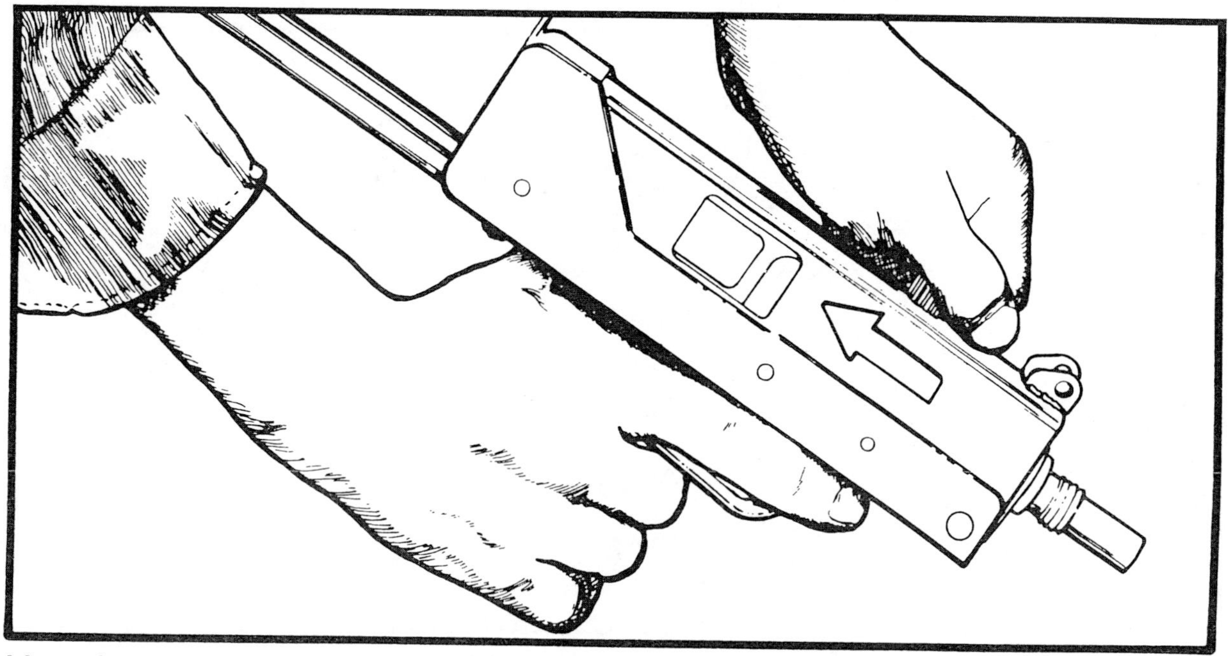

Move frame safety lever to "fire" position. Holding the gun by the pistol grip in the right hand and keeping the index finger outside the trigger guard pull back on the cocking handle to the rear with the left hand until the bolt is held to the rear by the sear. Move safety lever to "safe" position.

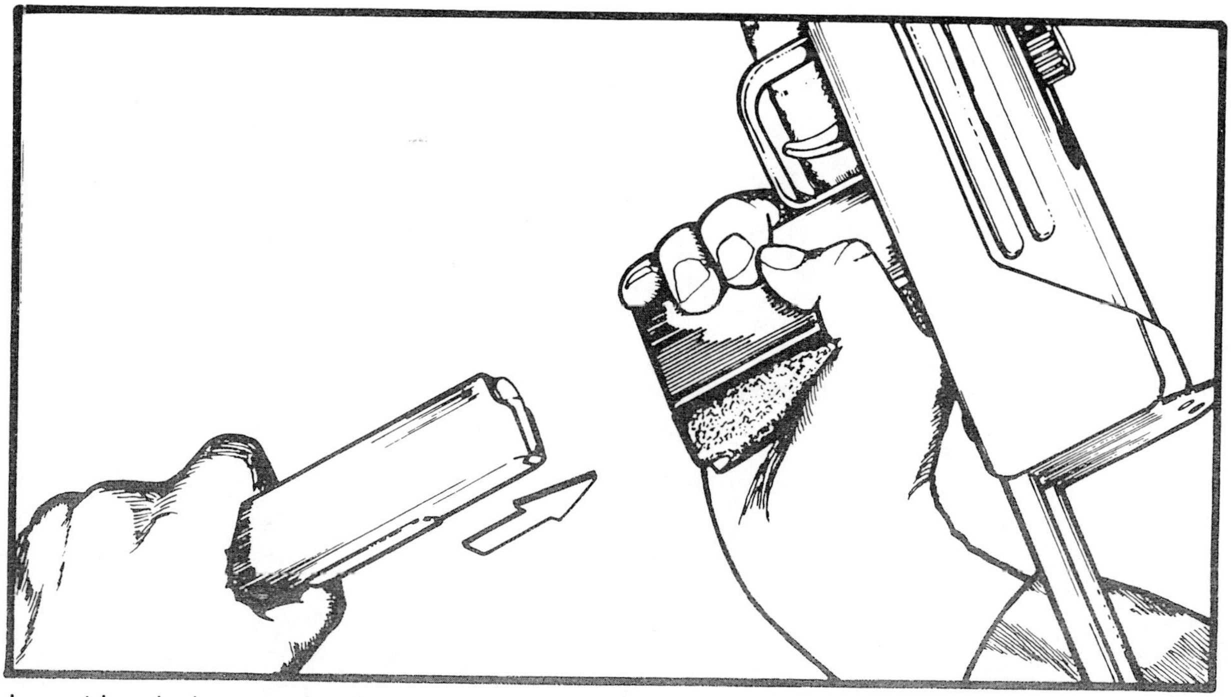

Insert loaded magazine in its housing until the magazine catch engages magazines. Select "semi" or "auto" fire as required, move frame safety to "fire" position. The gun is now ready to fire.

PLACING THE WEAPON ON "SAFE"

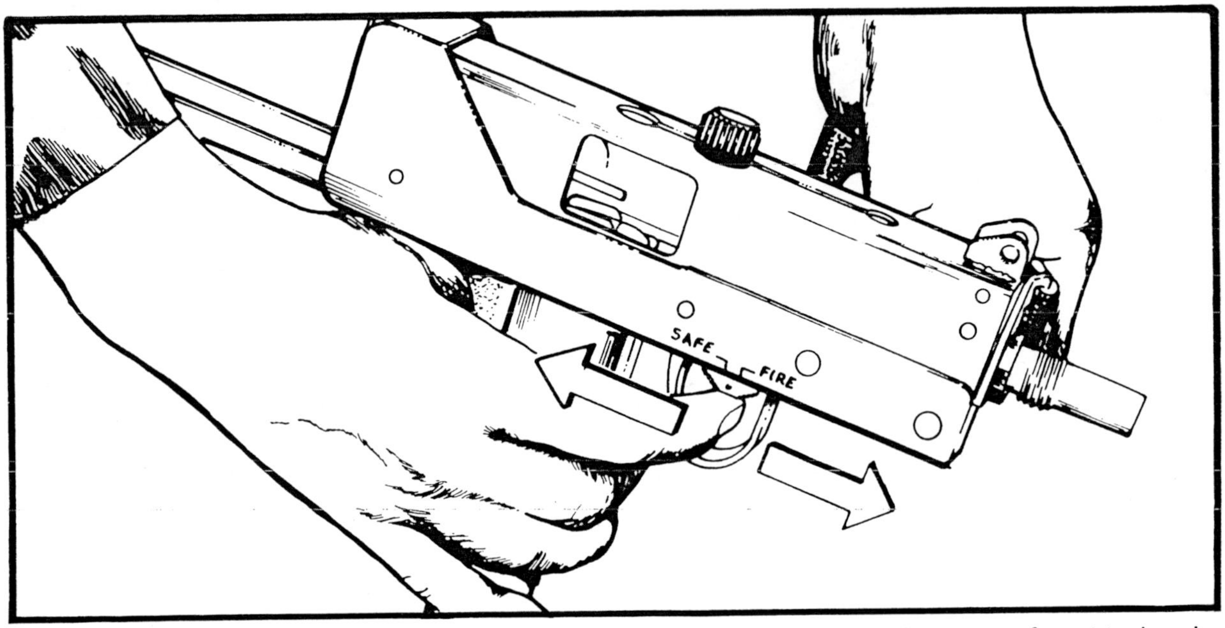

The safety is located on the right of the underside of the frame, it moves front to back, front is the "fire" position and back is the "safe" position.

REMOVAL OF MAGAZINE

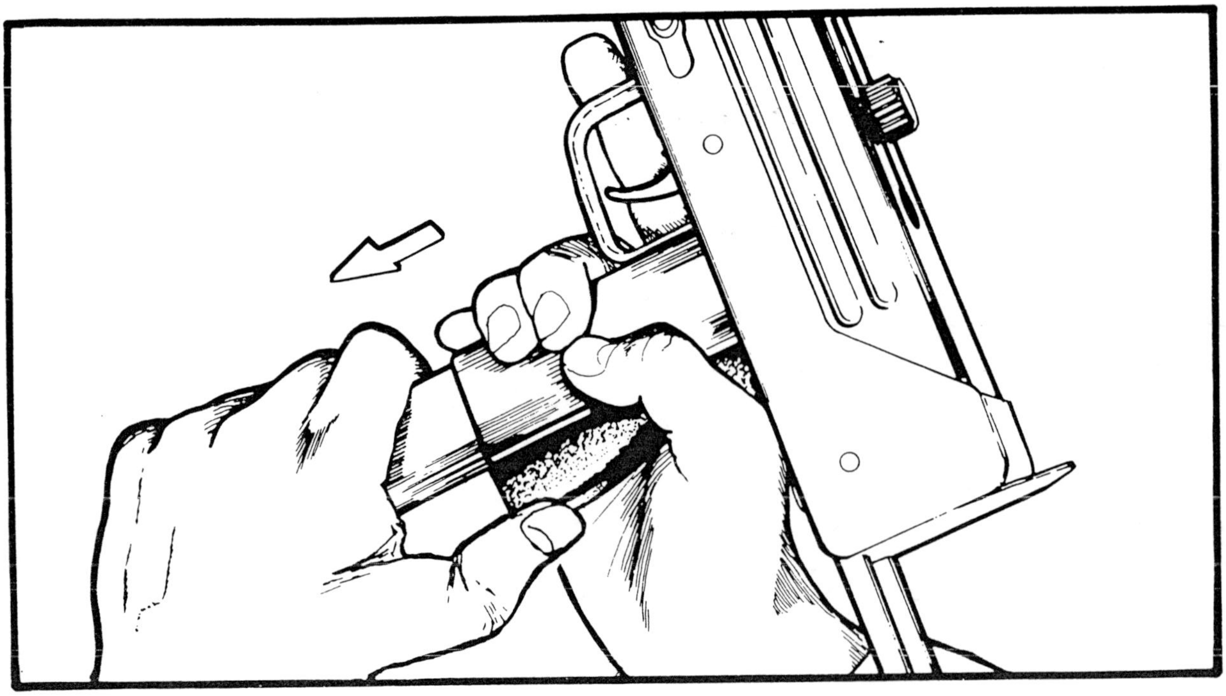

Grasp magazine in left hand and with left thumb press magazine catch to release magazine.

FIRE SELECTOR LEVER OPERATION

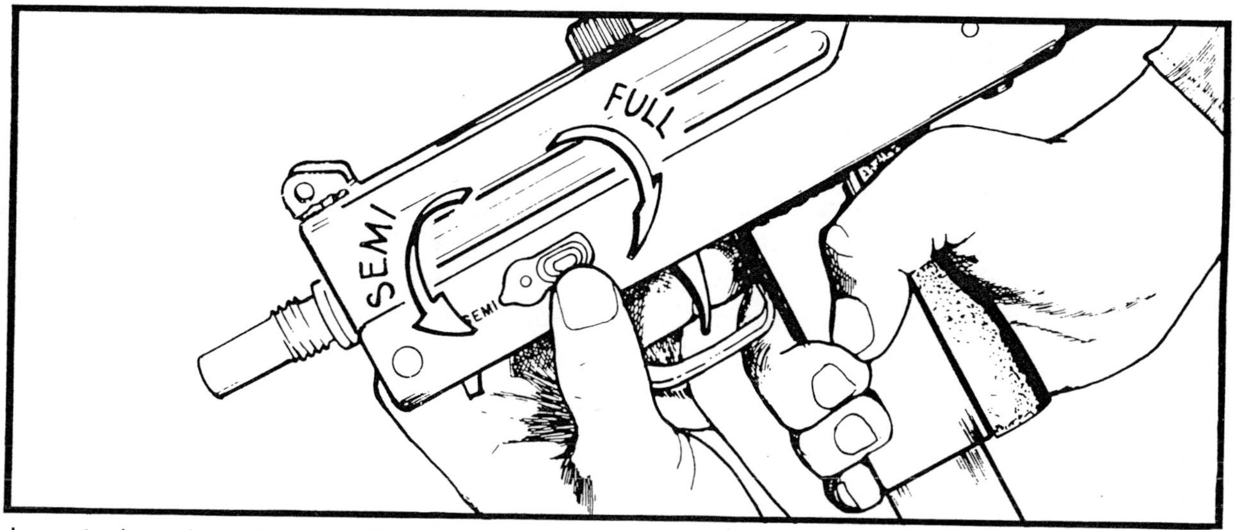

Located on the left side of the frame the selector is rotated to select "semi" automatic or "full" automatic fire.

Firing the M-10 & M-11 Submachine Gun

As the bolt is moved back to the cocked position, the recoil spring is compressed and the sear engages the sear notch of the bolt. When the trigger is pressed, the sear releases the bolt, which is driven forward by the recoil spring. During this forward movement, the bolt strips a cartridge from the magazine into the chamber. The bolt continues forward and fires the cartridge. When the cartridge is fired, the chamber pressure forces the bullet out of the muzzle of the barrel. At the same time, this pressure overcomes the forward movement of the bolt and starts it to the rear. By the time the bolt and empty case have moved to the rear far enough to open the rear of the chamber, the bullet has left the barrel, and the chamber pressure has diminished. (In the submachine gun, the chamber pressure is relatively low and the bolt is relatively heavy; this eliminates the need for positive locking and unlocking.) During the rearward movement of the bolt, the empty cartridge case is extracted and ejected, the recoil spring is compressed and the top round in the magazine moves up against the lips of the magazine. The rearward movement of the bolt is stopped by contact with the buffer plate.

Malfunctions

Malfunctions are usually the result of worn parts or improper care of the gun. A knowledge of how the gun functions enables the user to classify and correct the malfunction. Listed below are the types of malfunctions which might occur.

1. Failure to Feed. The top cartridge in the magazine is not positioned up and in

front of the bolt. Most malfunctions of the submachine gun are failures to feed caused by a defective or dirty magazine.

2. Failure to Chamber. The top cartridge from the magazine is not seated in the chamber.

3. Failure to Fire. The cartridge is chambered but does not fire.

4. Failure to Extract. If the cartridge fires, the chamber pressure will normally push the empty cartridge case out of the chamber. If the cartridge case is not completely removed from the chamber and the bolt is retracted, then there is a failure to extract. This malfunction seldom occurs.

5. Failure to Eject. The empty cartridge case is not ejected from the weapon.

6. Failure to Cock. If the bolt is retracted and is not held by the sear, or if, during firing, the bolt does not move to the rear far enough to clear the top cartridge in the magazine, the gun fails to cock.

Common Malfunctions. The two most common malfunctions are:

1. Failure to feed—usually caused by a defective magazine.

2. Failure to fire—usually caused by defective ammunition.

Causes of Malfunctions. The following chart lists common causes of various malfunctions and corrective action.

Prevention of Malfunctions. Periodic inspection and proper care and cleaning will reduce the possibility of the submachine gun malfunctioning.

Malfunctions	Cause	Corrective Action
Failure to feed.	Dirty or dented magazine. Weak or broken magazine spring. Worn magazine catch. Corroded ammunition. Worn or broken magazine catch.	Replace magazine. Replace magazine. Replace magazine. Replace ammunition. Replace magazine catch.
Failure to chamber.	Dirty chamber. Obstruction in chamber. Weak recoil springs.	Clean chamber. Remove obstruction. Replace recoil spring.
Failure to fire.	Defective ammunition. Defective firing pin. Weak recoil springs.	Replace ammunition. Replace bolt. Replace recoil springs.
Failure to extract.	Broken extractor.	Replace extractor.
Failure to eject.	Broken ejector. Broken or missing extractor.	Replace ejector. Replace extractor.
Failure to cock.	Worn sear. Worn sear notch. Bent guide rods. Low powered ammunition.	Replace sear. Replace bolt. Straighten. Replace ammunition.

STRIPPING AND REASSEMBLING

The Model 10 and Model 11 have been designed so that no special tools are necessary in order to strip or reassemble the weapon.

Stripping

Before starting to strip the weapon remove the magazine and check the barrel chamber by looking through the EJECTION PORT to make sure there are no live rounds in the weapon.

Push receiver pin catch back and remove receiver pin. On alternate configuration where there is no receiver pin catch, a sharp push on the receiver pin is all that is required for removal.

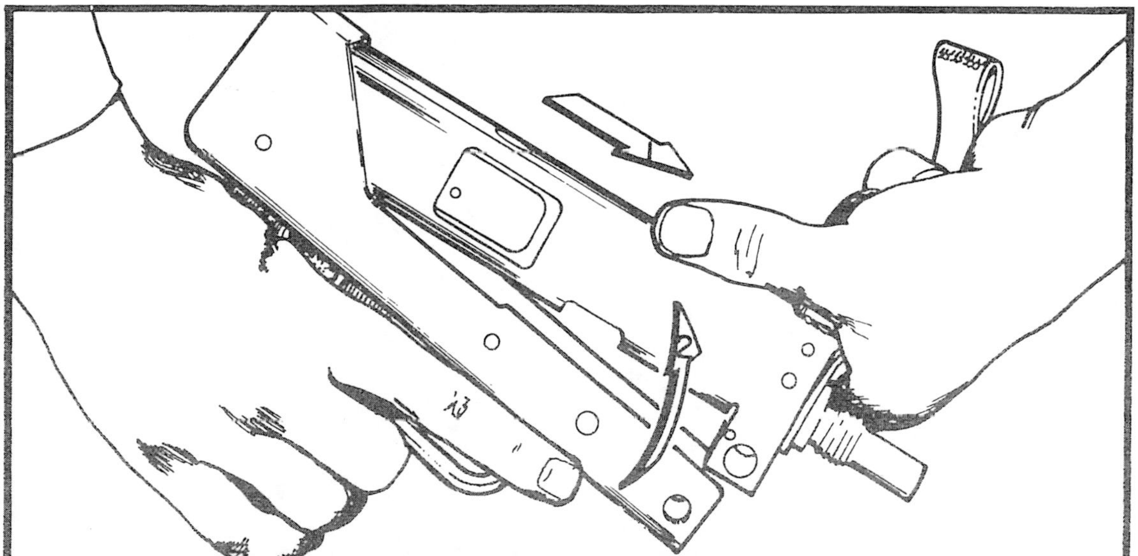

Push upper receiver from frame.

Slide cocking handle to rear of guide slot, rotate handle a few degrees and pull from bolt. The bolt and recoil spring assembly can now be removed from the rear of the receiver.

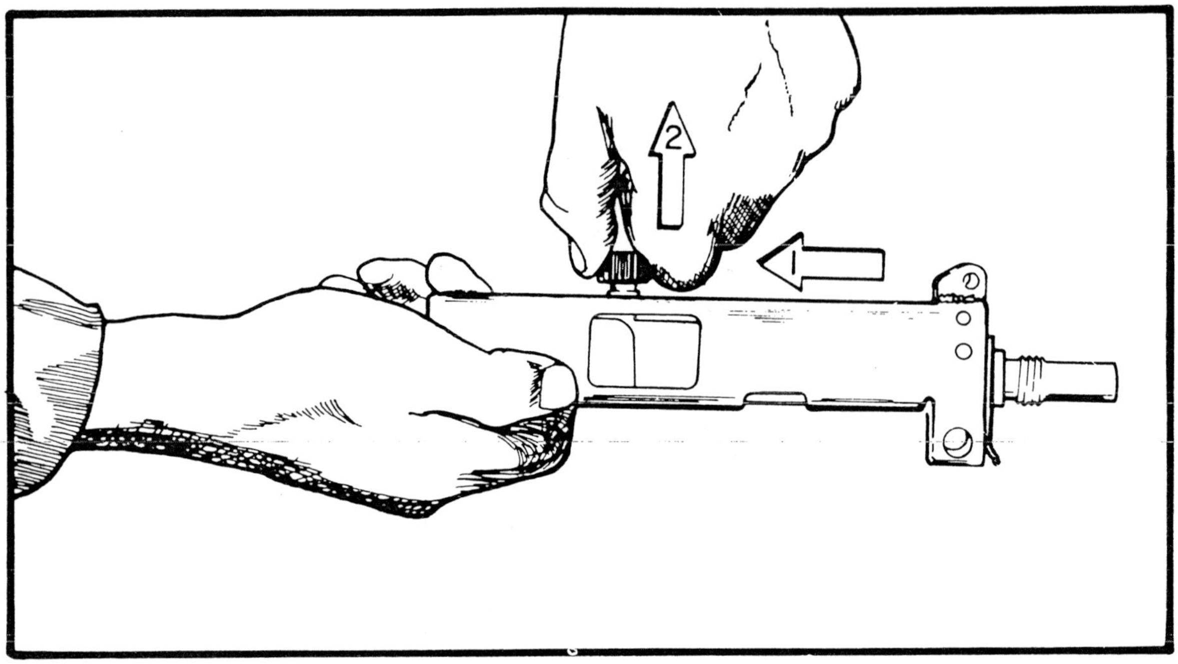

Removing cocking handle.

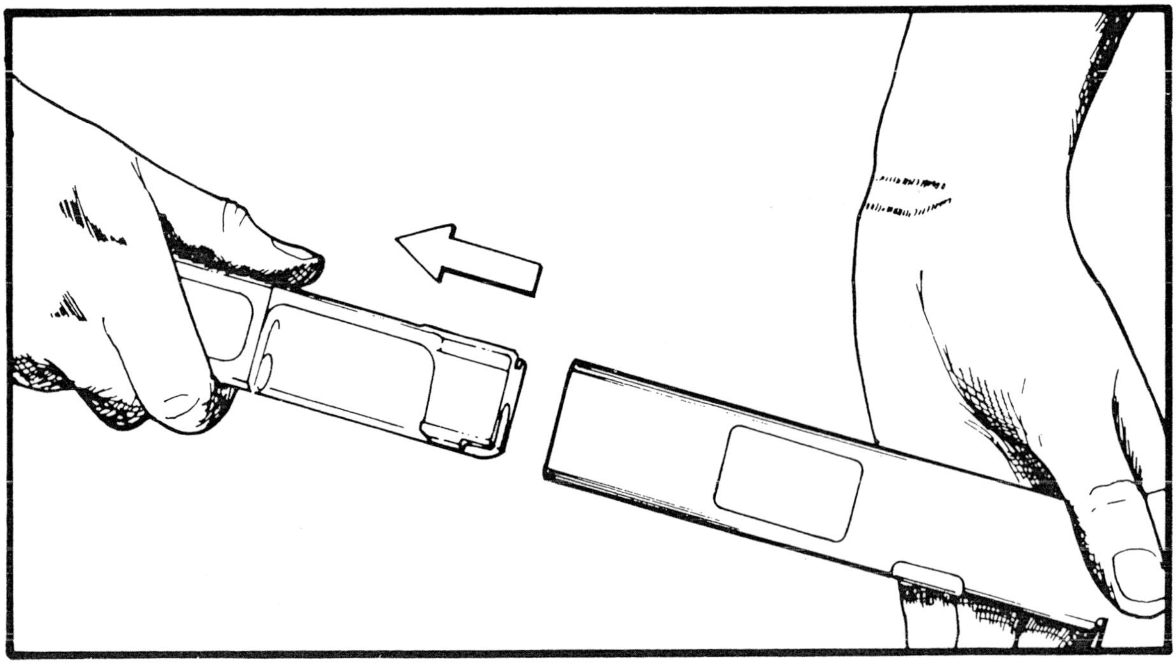

Bolt remover from rear of receiver.

The procedure mentioned above is generally sufficient for normal maintenance and cleaning. If necessary to clean trigger mechanism, removal is as follows:

Trigger Mechanism

Lift retainer and remove selector lever this also allows removal of sear, sear spring and tripping lever. Push trigger pin out of the left side for removal of trigger or torsion spring. Note: for reassembly, free arm of torsion spring is located on rear side of disconnector. Disconnector may be replaced by using a pin punch to remove spring pin.

Extractor

Use pin punch to remove extractor spring pin.

Recoil Spring

Compress spring and move guide rod clear of bolt and use pin punch to remove spring pin from guide rod.

Receiver Pin Catch

Use pin punch to remove spring pin.

Magazine Catch

Use flat bladed screwdriver to remove screw in hand grip for access to catch.

Safety Catch

Use pin punch to remove catch spring pin.

Stock Latch

Press down lightly at center of pin and slide to either side to remove retaining pin. Invert frame and latch will drop out.

Magazine

M10/9mm Para, M11 .380 ACP

Depress stud in floor plate and slide plate from magazine case. Keeping finger over the bottom of the magazine to prevent the magazine spring from flying out. Remove the magazine spring and the magazine follower.

M10/.45 ACP

Lift the tab in the floor plate by inserting a screwdriver in the hole. Remove base plate, placing the finger over the bottom of the magazine to prevent the magazine spring from flying out.

Remove the magazine spring and the magazine follower.

NOTE: To achieve the proper functioning when reassembling the M11 and M10/.45 magazine ensure that the slope of the top coil of the spring corresponds with the slope on the follower.

REASSEMBLY

Items are reassembled in the reverse order to the stripping procedure.

CLEANING AND MAINTENANCE

It is essential for reliable operation and performance that the weapon receive careful maintenance. It should be cleaned at the end of each day's firing.

The Gun need only be "field-stripped" for this maintenance.

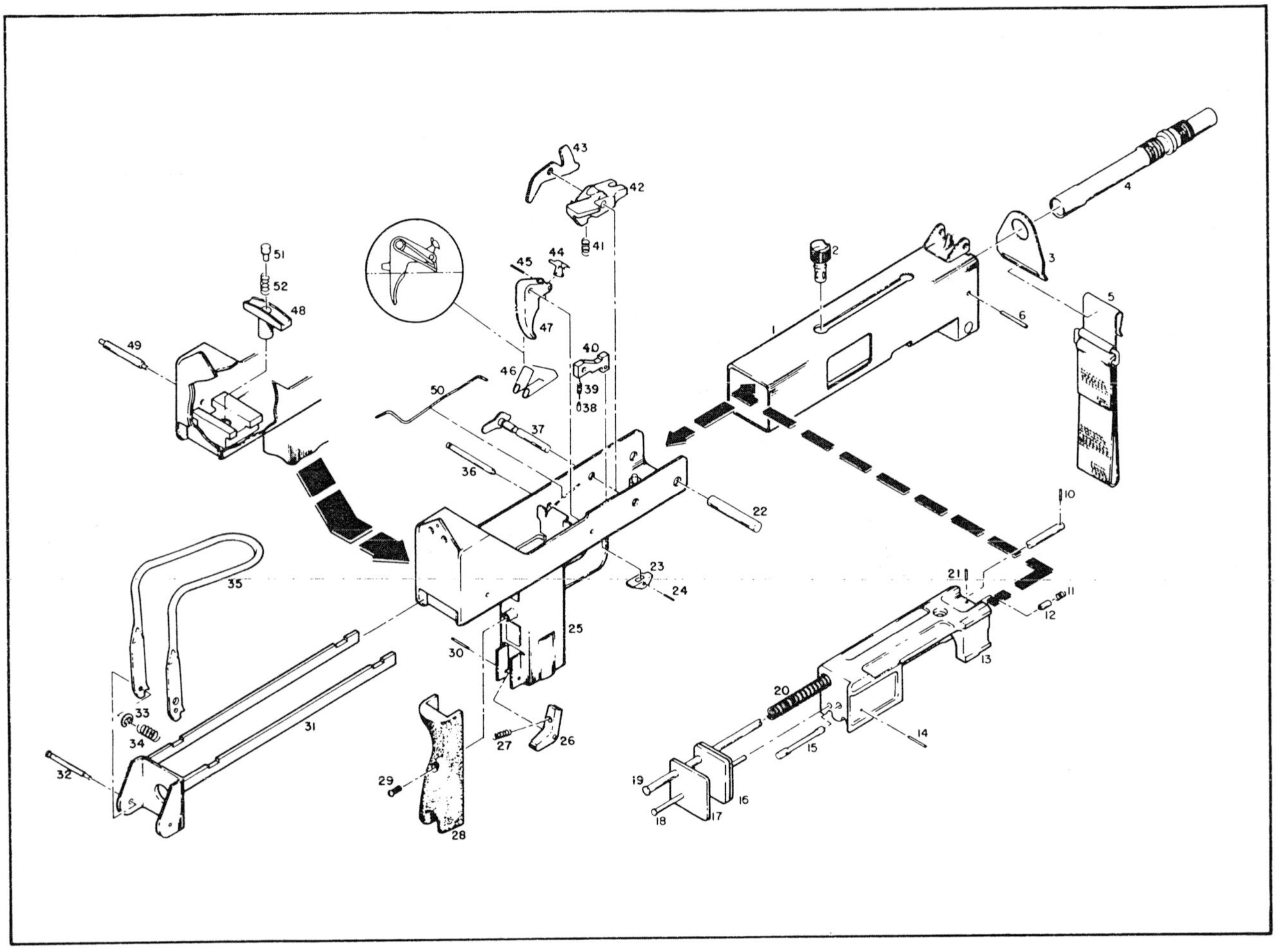

M10/.45 ACP

1 Receiver
2 Bolt Handle
3 Hanger
4 Barrel
5 Strap Assembly
6 Spring Pin, Barrel
10 Spring Pin, Recoil Spring Rod
11 Compression Spring, Bolt Handle Detent
12 Bolt Handle Detent
13 Bolt
14 Spring Pin, Extractor
15 Extractor
16 Buffer
17 Plate
18 Ejector
19 Guide Rod
20 Compression Spring, Recoil
21 Spring Pin, Bolt Handle Detent
22 Receiver Pin
23 Button
24 Spring Pin, Safety
25 Frame Assembly
26 Magazine Catch
27 Compression Spring, Magazine Catch
28 Grip
29 Grip Screw
30 Magazine Catch Pin
31 Stock Rail
32 Stock Hinge Pin
33 Retaining Ring
34 Compression Spring, Stock Hinge
35 Wire Form Butt
36 Trigger Pin
37 Sear Pin
38 Safety Detent Plunger
39 Compression Spring, Safety Detent
40 Safety Slide
41 Compression Spring, Sear
42 Sear
43 Trip
44 Disconnector
45 Spring Pin, Disconnector
46 Torsion Spring, Trigger
47 Trigger
48 Latch
49 Stock Latch Pin
50 Retainer
51 Plunger, Stock Latch
52 Compression Spring, Stock Latch

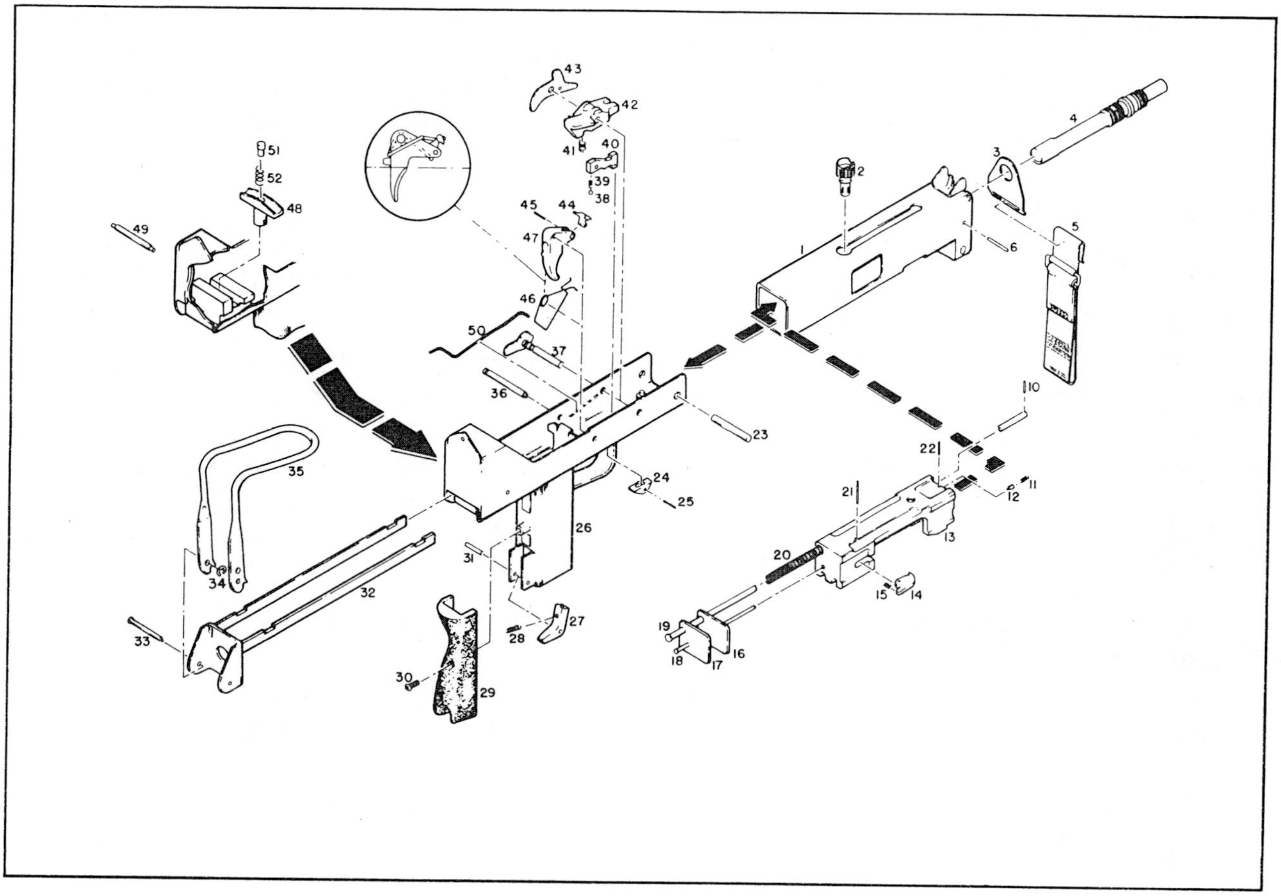

M11/.380 ACP

1 Receiver
2 Bolt Handle
3 Hanger
4 Barrel
5 Strap Assembly
6 Spring Pin, Barrel
10 Spring Pin, Recoil Spring Rod
11 Compression Spring, Bolt Handle Detent
12 Bolt Handle Detent
13 Bolt
14 Extractor
15 Compression Spring, Extractor
16 Buffer
17 Plate
18 Ejector
19 Guide Rod
20 Compression Spring, Recoil
21 Spring Pin, Extractor
22 Spring Pin, Bolt Handle Detent
23 Receiver Pin
24 Button
25 Spring Pin Safety
26 Frame Assembly
27 Magazine Catch
28 Compression Spring, Magazine Catch
29 Grip
30 Grip Screw
31 Magazine Catch Pin
32 Stock Rail
33 Stock Hinge Pin
34 Retaining Ring
35 Wire Form Butt
36 Trigger Pin
37 Sear Pin
38 Safety Detent Plunger
39 Compression Spring, Safety Detent
40 Safety Slide
41 Compression Spring, Sear
42 Sear
43 Trip
44 Disconnector
45 Spring Pin, Disconnector
46 Torsion Spring, Trigger
47 Trigger
48 Latch
49 Stock Latch Pin
50 Retainer
51 Plunger, Stock Latch
52 Compression Spring, Stock Latch

Barrel and Receiver Group Cleaning

A. Use cleaning rod, patch and solvent. Stubborn residue can be removed with a bristle bore brush.

B. Swab bore with patch saturated in solvent.

C. Use solvent saturated patch and swab inside of receiver to remove residue.

D. Use clean dry patch to dry barrel and inside of receiver group.

E. Dampen patch with light oil and swab barrel and receiver.

Bolt Assembly Cleaning

A. Remove powder residue with solvent and patch.

B. Wipe all surfaces of bolt to remove carbon. The bolt face should be completely free of carbon and other residue.

C. Wipe dry and apply a light coat of oil.

Frame Assembly Group Cleaning

A. Remove residue build-up on exposed surfaces with solvent-soaked cloth.

B. Clean around the lockwork. The weapon is designed to tolerate some dirt in the frame group but excessive dirt will impair operation. Detail cleaning is required after prolonged operation or improper functioning of weapon.

C. Wipe dry and use oily patch to coat exposed metal areas with light film of oil.

Magazine Cleaning

A. Inspect magazine for damaged areas. Damaged magazines should be discarded.

B. Clean lip area and top of the follower.

C. Wipe with oily patch.

Suppressor Cleaning

The suppressor requires no other maintenance apart from ensuring that the bore is kept clear of obstructions. Inspection must be made with the suppressor removed from the weapon. The front end cap of the suppressor is replaceable. It should be replaced after each 500 rounds of firing or sooner if the noise level has increased significantly.

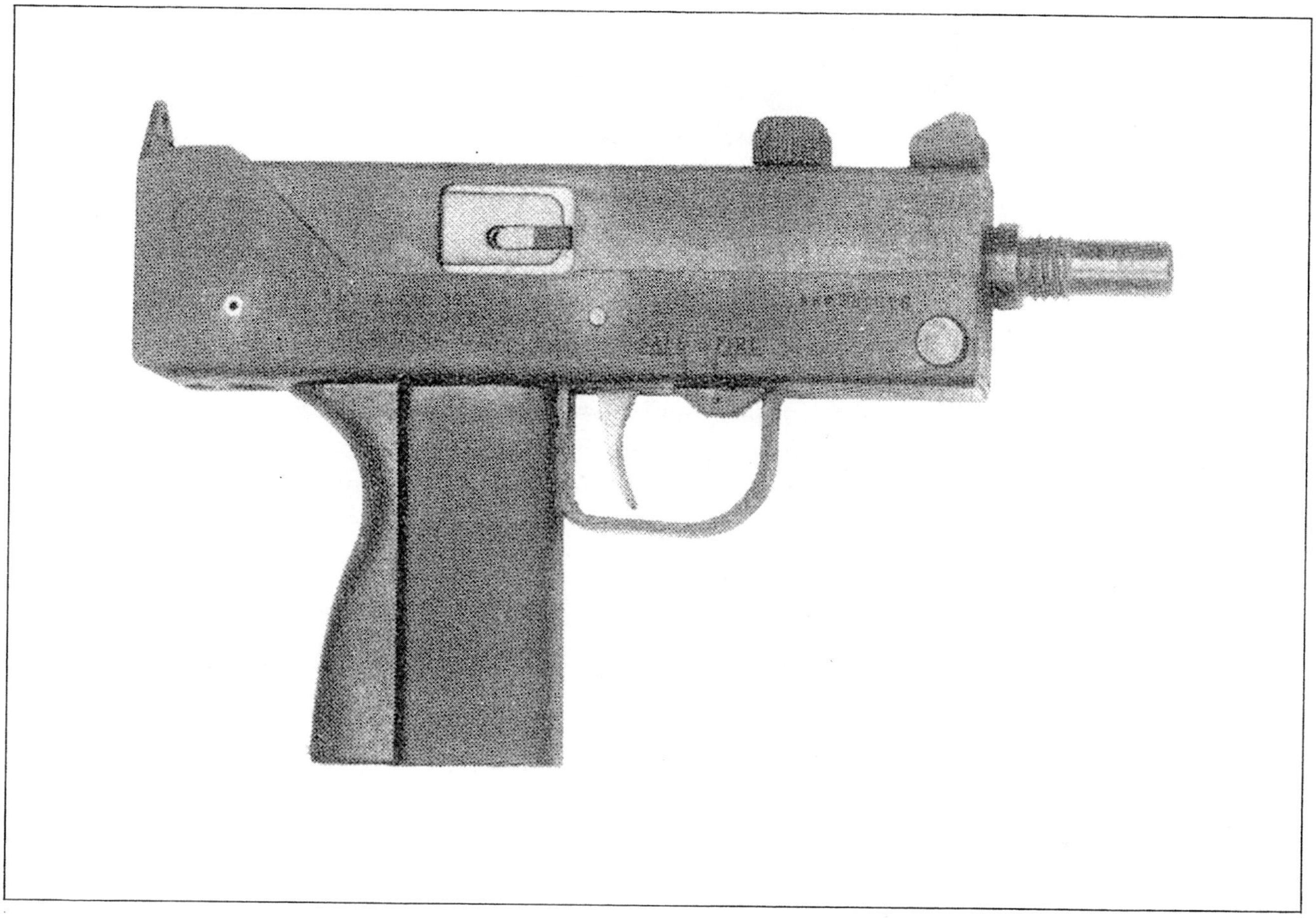

Instruction for use of Ingram Semiauto Pistols and Carbines

The semiauto Ingram pistols and carbines were designed from the Ingram submachine gun.

The major differences between the two guns consists of: (1) no provision in the new designed frame for a sear pin, selector, (2) no provisions for a stock in the pistols, and (3) trigger, sear, sear pin and trip are captive in the frame as to discourage anyone from trying to change this unit to full-auto fire.

As this is an open bolt firing design, these few simple instructions must be adhered to:

Loading—

1. Remove the magazine, move the safety to "fire" and pull bolt handle to the rear until the bolt locks on rear catch.
2. Move the safety to "safe" position.
3. Insert the magazine.
4. To fire, move the safety to "fire" position, point at target and pull the trigger. (This action strips a cartridge from the magazine, chambers, fires, ejects and recoils to open position.)

To Clear Jams—

1. Point the gun toward a safe place only, pull the bolt handle to full cock, move the safety to "safe" position, remove the magazine and clear the jam.

VARIOUS INGRAM CONFIGURATIONS THAT WERE AVAILABLE

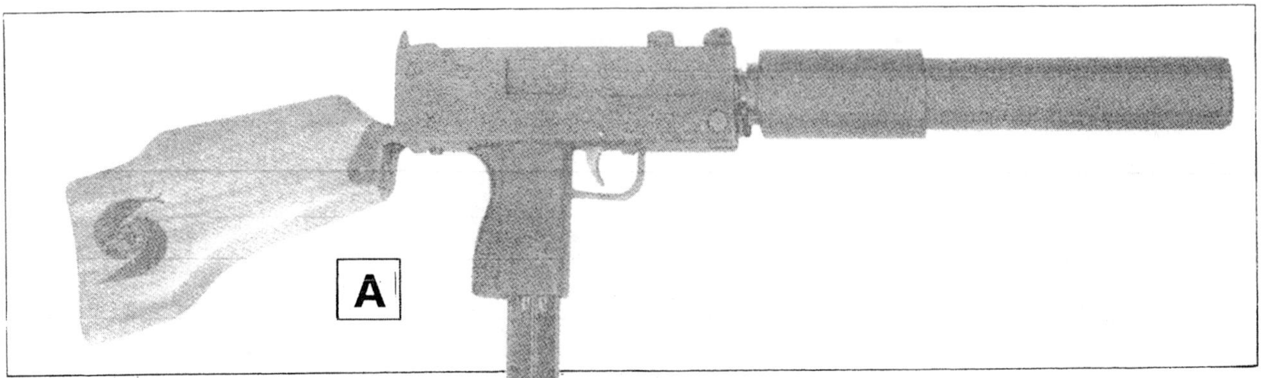

Deluxe Ingram M-10

A. M-10, in 9mm, with wooden buttstock, 32-round magazine and Sionic type sound suppressor. The basic difference in this arrangement over the conventional design is the substitution of the wire form stock with the removable deluxe wood stock.

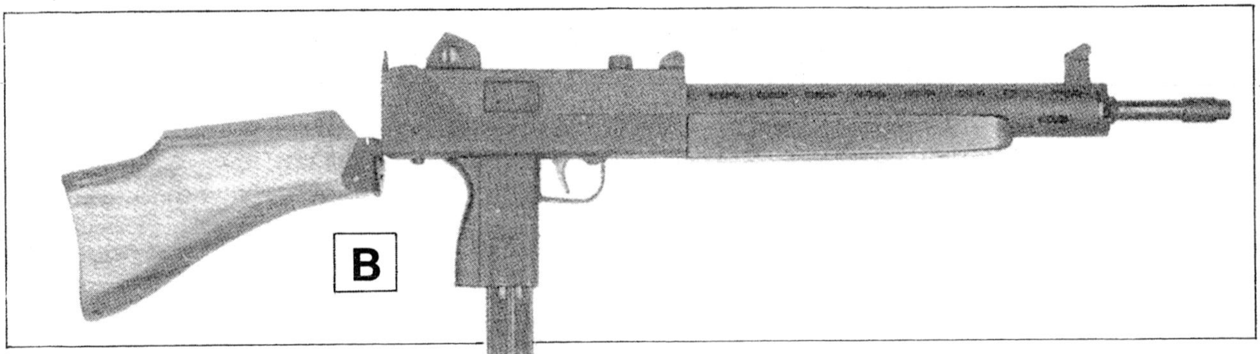

Deluxe Carbine

B. M-10, in 9mm, with deluxe LR carbine barrel/receiver and wooden buttstock. These LR carbine configurations are designed to give greater accuracy and more stability while the gun is being fired. These LR barrel/receivers can also be installed on the SM-10 and SM-11 models with the same results. Please note the addition of the larger rear sight.

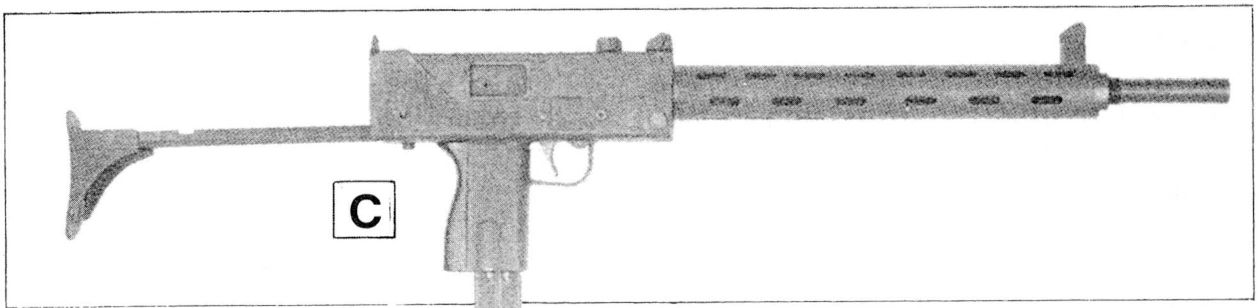

Standard Carbine

C. M-10, in 9mm, virtually the same specifications as the Deluxe Carbine, with the exception of the wooden forearm and deluxe wooden stock. The same fine quality goes into the manufacture of these standard LR receivers as does any of R.P.B.'s products.

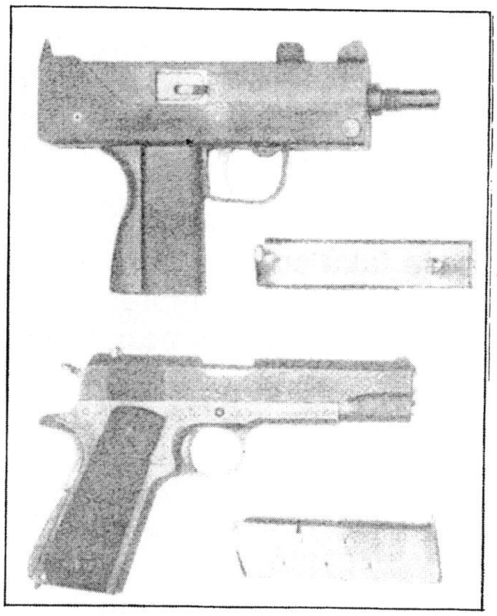

With regard to size . . .

Shown here, for comparison, we have photographically illustrated an SM-11 .380 INGRAM and a government model .45 automatic. You will note that the two guns are very close to the same size. However, the SM-11 offers over twice the firepower, with its 16-round magazine, than that of the .45 auto. For this reason, the SM-10 or SM-11, semiautomatic, makes an excellent home defense weapon or just a pleasant change of pace for the firearms buff.

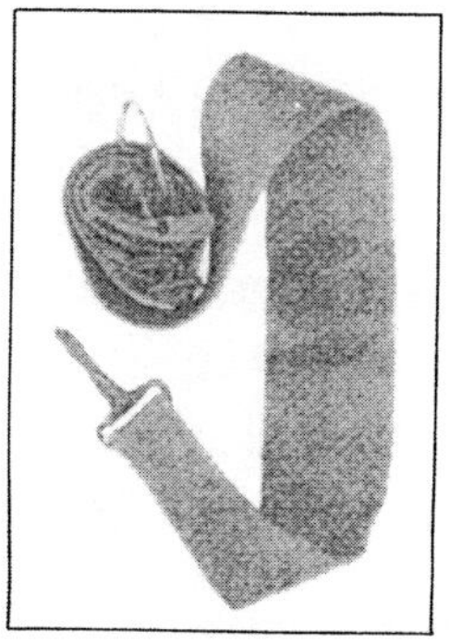

The Ingram Alternate Shoulder Sling

These durable slings are constructed of heavy, O.D. color, web material and fully adjustable for your carrying convenience. Shown below is the M-10 fitted with the deluxe buttstock and illustrating one way the shoulder sling may be incorporated with the INGRAM, or any other similar firearm. A must for every INGRAM owner.

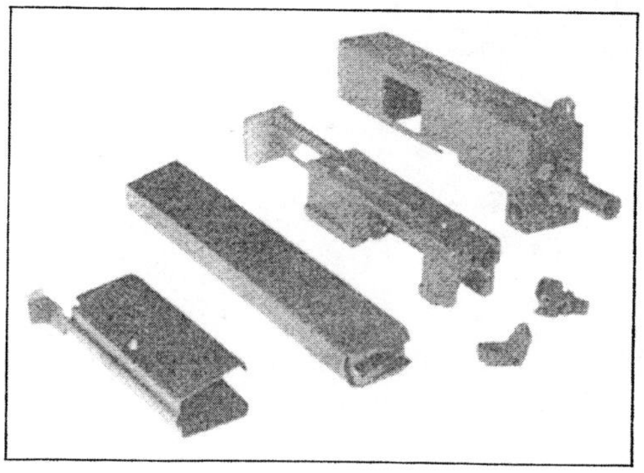

Ingram M-10 Conversion Unit

With the possibility of the U.S. government adoption of the 9mm Para as its standard combat pistol cartridge, the feasibility of owning a .45 INGRAM conversion to 9mm kit could be more of a necessity than a luxury. These kits come complete and very easy to install in your .45 cal. M-10 or SM-10.

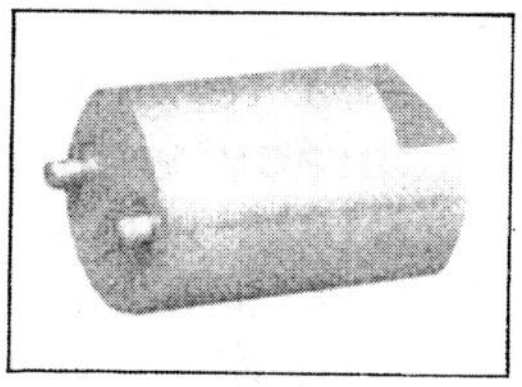

Ingram Suppressor Disassembly Tool

These disassembly tools are the only practical way of dismantling an INGRAM sionics type sound suppressor. Made of solid steel construction for long reliable life.

Sionic Type Sound Suppressor Wipes

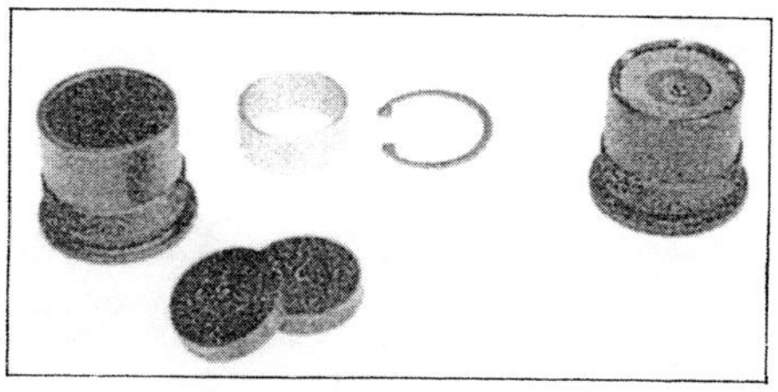

These are our new end caps for the INGRAM suppressors, with replaceable urethane disks. These units come complete with wipe housing, spacer, snap ring, snap ring washer and two sets of replacement urethane disks. Please state model and caliber. Extra sets of urethane disks also available.

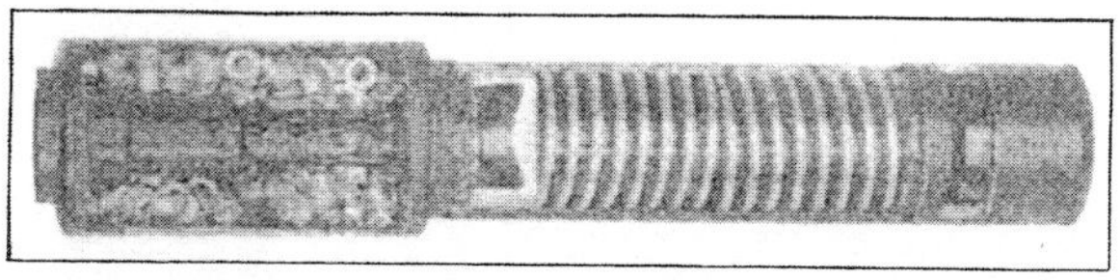

The Ingram Suppressor

Our suppressors have been said to be the best in the world. Specially designed for years of service, through the use of modern technology.

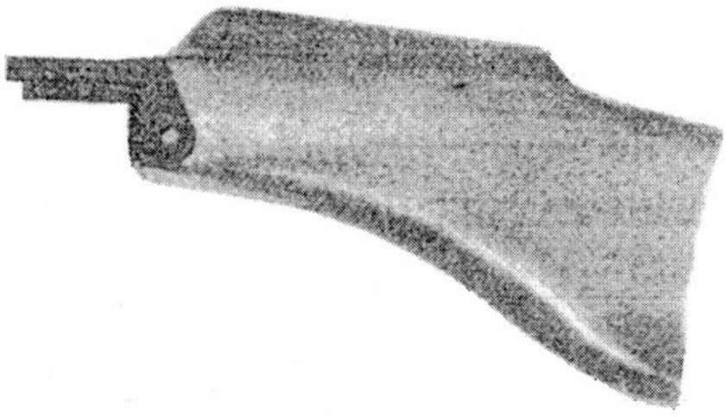

The Ingram Deluxe Wooden Buttstock

The wooden buttstock is designed to give you greater accuracy while firing your INGRAM in the full-auto or semiauto mode. These stocks also give an added touch of appeal.

M-10/M-11 and SM-10/SM-11 Fake Suppressors

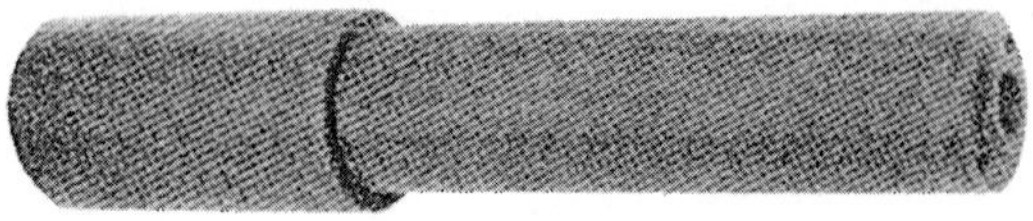

A barrel extension designed to simulate, in appearance, our popular sound suppressor. These fake suppressors aid in accuracy on either the full-automatic or semiautomatic version of the INGRAM. Suppressor kits cannot be installed in these extensions.

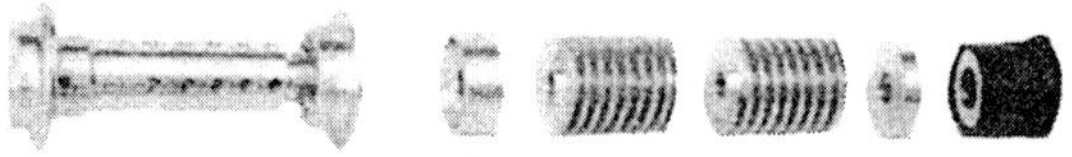

Ingram Suppressor Parts Kit

These two replacement parts for the M-10, M-11, SM-10 and SM-11 suppressors. Everything necessary to restore your old INGRAM suppressor to new condition. Complete package, no tubes included.

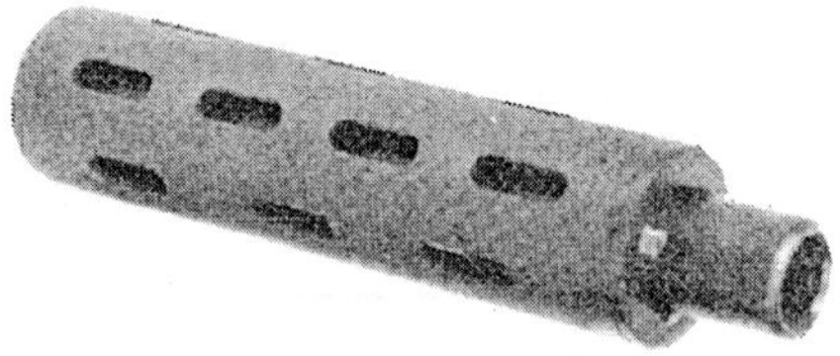

The Ingram M-10/SM-10 Alternate Barrel Extension

Constructed of heavy steel tubing, these extensions are a great alternate aid in maintaining accuracy. Vented for more efficient cooling while gun is in use, they also contribute to the appearance of a traditional submachine gun.

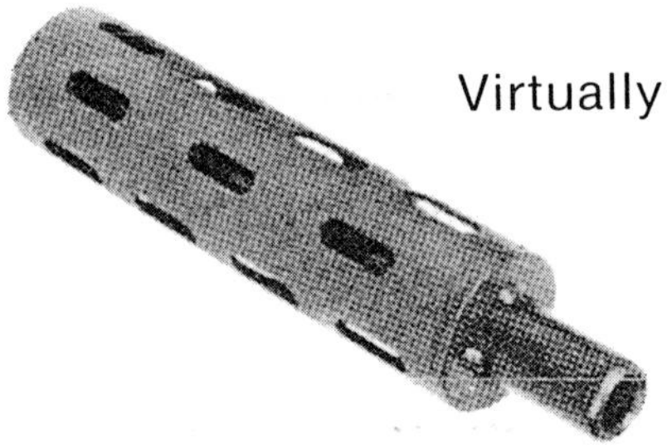

Ingram M-11/SM-11 Alternate Barrel Extension

Virtually the same extension as for the M-10/SM-10, only threaded for the M-11/SM-11 barrel. Like all barrel extensions, these are dull, nonreflective finished and match the finish of the INGRAM.

10" Barrel with Upper Receiver

Increase your INGRAM's accuracy and velocity with one of these extended barrel/receiver assemblies. Originally designed for secret special forces units, these barrel/receivers are now available to the general public.

Wooden Stock Holster

Carry your wooden buttstock, readily accessible, on your hip. The handy spring steel belt clip, on the back side, makes for easy belt adaptation and removal. The quick release Velcro fastened strap holds the stock securely in place while remaining active. For M-10 and M-11.

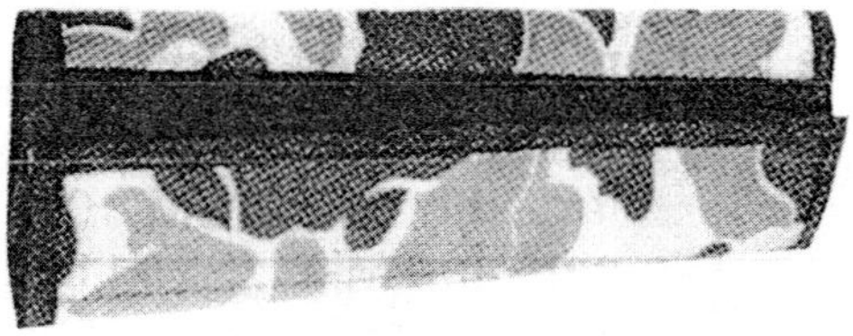

Nomex, Suppressor and Fake Suppressor, Cover

Made of long-lasting nylon material and available in black, O.D., green camo and desert camo. The Velcro strip provides easy fastening or removal of this cover. Also available are the rubber suppressor covers which are not illustrated in this manual.

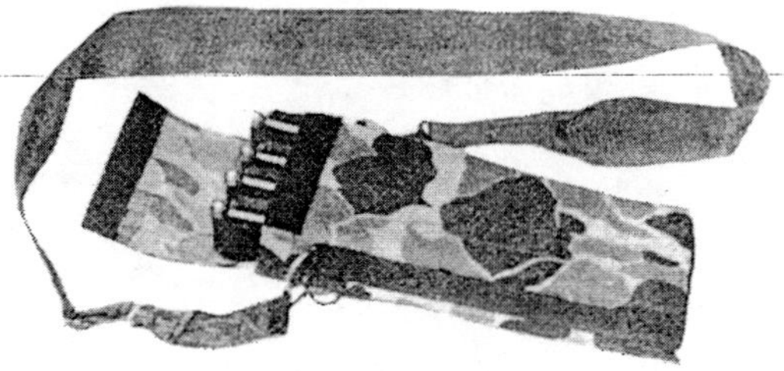

M-10 .45 Cal. Ingram 40-Round Four-Cell Pouch

These 40-round magazines and four-cell pouches allow you to carry an extra 160-round of .45 cal. ammunition for your M-10 or SM-10. The pouch, with sling, is made of durable nylon with a Velcro fastener on the flap.

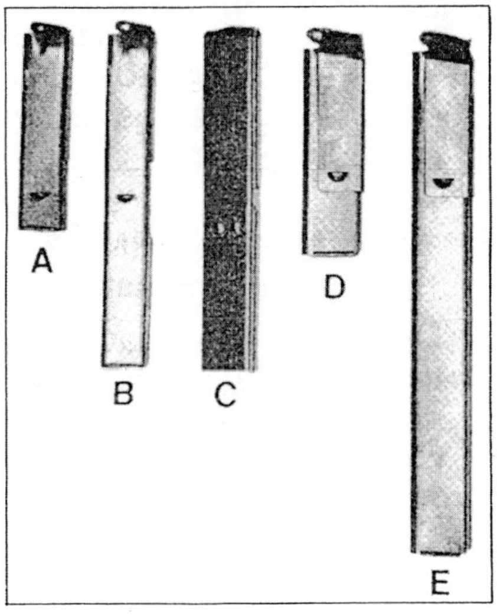

Our Leather Suppressor Holster

Carry your suppressor or fake suppressor on your hip and ready for instant service. Our suppressor holsters are constructed of fine heavy-stitched leather with a Velcro securing flap for easy access. Fitted with a leather belt loop that will accommodate most wide belts. Will also accept barrel extensions.

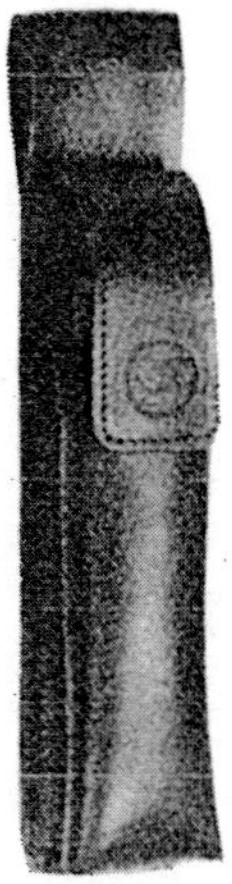

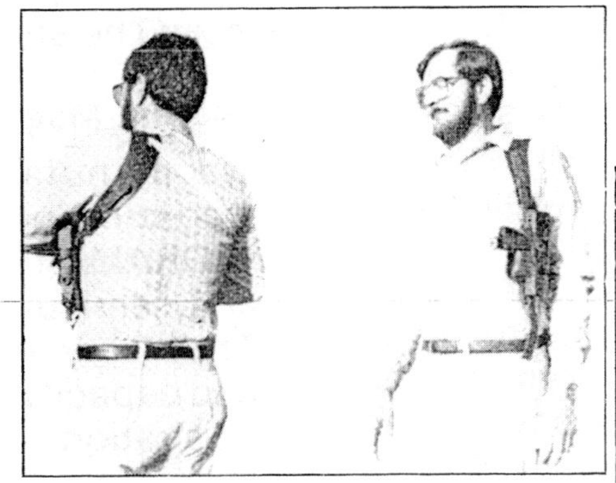

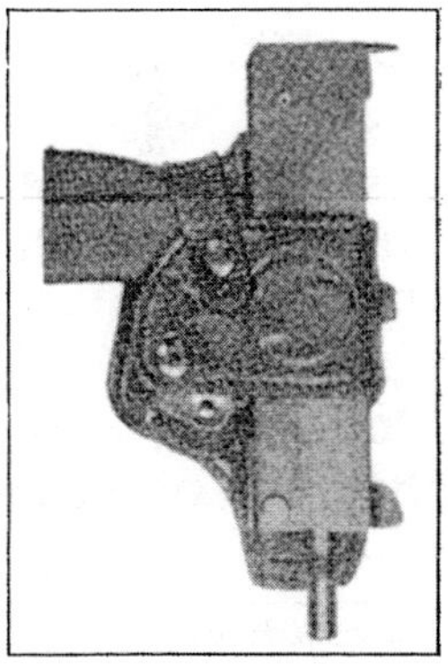

The M-11 Convertible w/Shoulder Harness

Wear your M-11 or SM-11 under your arm or on your hip with the utmost convenience. Made of high quality leather with the "COBRAY" logo stamped on the face of the holster. Quick, easy snap release.

Our shoulder harness is designed to be used with the M-11 convertible. Fully adjustable for any size shoulders. An ideal way to carry your INGRAM, concealed, under your coat or jacket.

Various Magazines for the M-10/SM-10, M-11/SM-11

A. The M-11/SM-11 .380 16-round magazine. Smaller, more compact, and very concealable.

B. The M-11/SM-11 .380 32-round magazine. When extreme firepower is the key.

C. The M-10/SM-10 9mm 32-round magazine. May also be used in our conversion units.

D. The M-10/SM-10 .45 cal. 10-round magazine. Ideal for semiautomatic fire when accuracy is more important.

E. The M-10/SM-10 .45 cal. 40-round magazine. Ultimate in INGRAM firepower.

Kevlar Inserts

This material was developed by one of the country's leading chemical companies as a tough fiber-like substance which, when layered, will provide a bullet-proof panel. Most of the flak vests on the market today use this material in forming their hard, nonpliable body armor. Notice that we leave ours soft for the sake of movement.

The Ingram Magazine Loading Tool

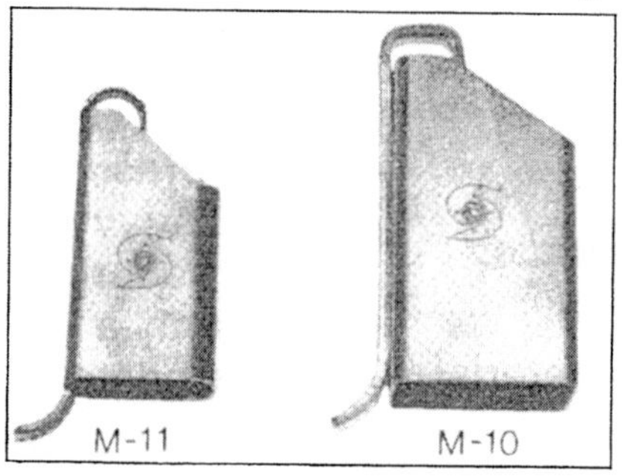

An essential tool for loading the M-10/SM-10 .45 cal., and M-11/SM-11 .380 cal. magazines. Compact and easy to carry, these loading tools are essential for ease of loading. Whether loading in the field or loading at home, these loading tools will prove to be a most valuable asset. (The 9mm does not require a loading tool.)

Ingram Magazine Pouches

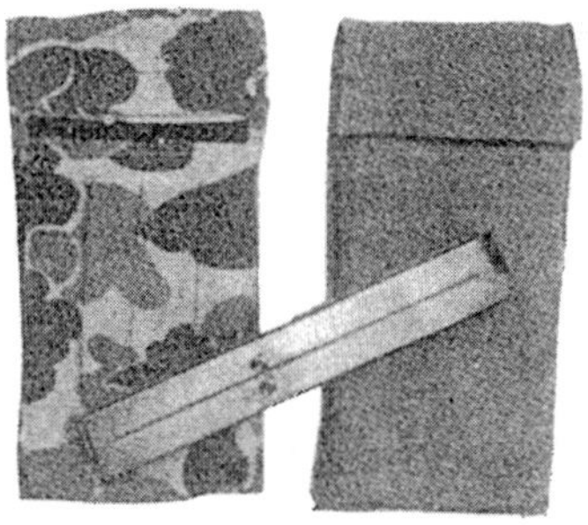

We manufacture various magazine pouches for the INGRAM submachine gun and semiautomatic pistols. They vary in size, caliber and capacity. Shown in this illustration are the three-cell and four-cell pouches. Available in either solid or camouflage patterns.

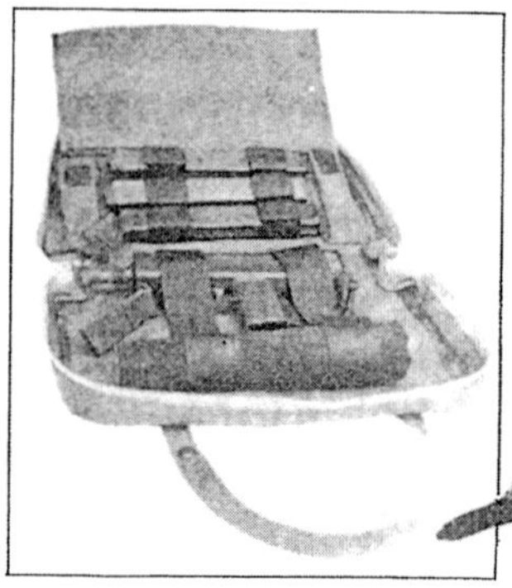

R.P.B. SWAT Case

For use with INGRAM guns. SWAT cases will accommodate one M-10 or M-11, three magazines, one suppressor or barrel extension, loading tool and cleaning accessories. The case is also a very effective bulletproof vest when Kevlar panels are inserted and the case worn in the manner shown.

The M-10 Leather Holster

What could be handier than carrying your M-10, in 9mm or .45 cal.. on your hip, ready for immediate action. Will accommodate the larger M-10 submachine gun or SM-10 semiautomatic pistol. Belt loops fit most larger size belts.

The M-11 and SM-11 Cut-Away Holster

Originally manufactured to fill special government needs. Quick release double snap makes it one of the fastest SMG holsters on the public market. Made of the same fine leathers as our other holsters. Available in black only.

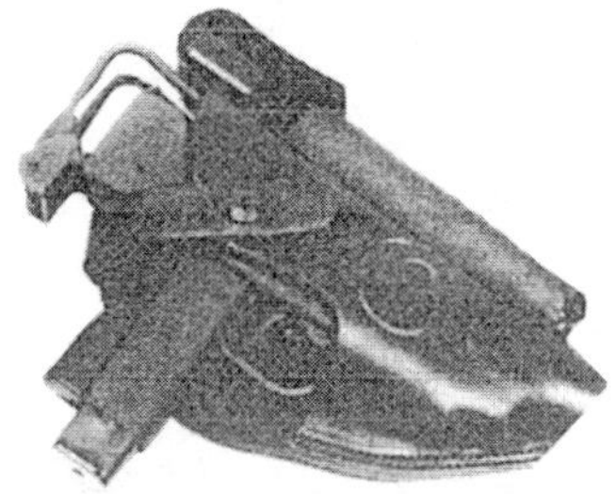

M-11 Black Leather Holster w/Mag. Pouch

Designed for hip use only and can be used with the M-11 or SM-11 INGRAM. Has the quick release snap, but will also carry an extra 32-round magazine.

R.P.B. M-11/SM-11 Operational Briefcase

An ordinary briefcase, containing the world's deadliest submachine gun. The trigger mechanism is extended through the bottom of the case, so the gun may be fired while still in the case. Must be used with suppressor or barrel extension. Classified by the BATF as an assassination device and must be federally registered.

The Ingram Carry-All

Sturdy 8-ounce cotton duck tote bag will hold your INGRAM, ammo and extra magazines. Larger version also available.

The Ingram T-Shirt

T-shirts are made of cotton and polyester in yellow and bear the COBRAY logo in red, blue and black. Behind logo is a line silhouette of the INGRAM SMG. In S, M, L, XL.

Cobray Key Fob

Black leather with COBRAY emblem deeply embossed in the center. Sturdy 1¼-inch ring.

Our Commando Wallet

Rugged nylon parachute cloth is water and mildew resistant. Five inner pockets. Has Velcro fastener.

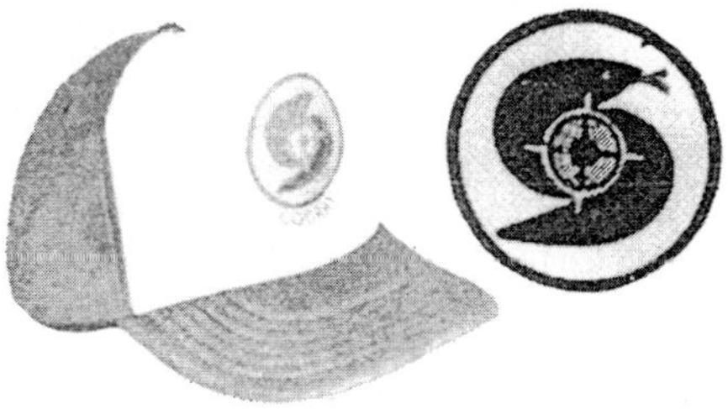

R.P.B. Cobray Hat and Shoulder Patch

Hat is black and white polyester. INGRAM logo is displayed, in full color, on the front. Adjustable back strap. COBRAY shoulder patch is embroidered in yellow, red, blue and black.

The Commando Watchband

Crystal cover protects when needed or folds out of the way. Nylon webbing will last forever. Velcro fasteners.

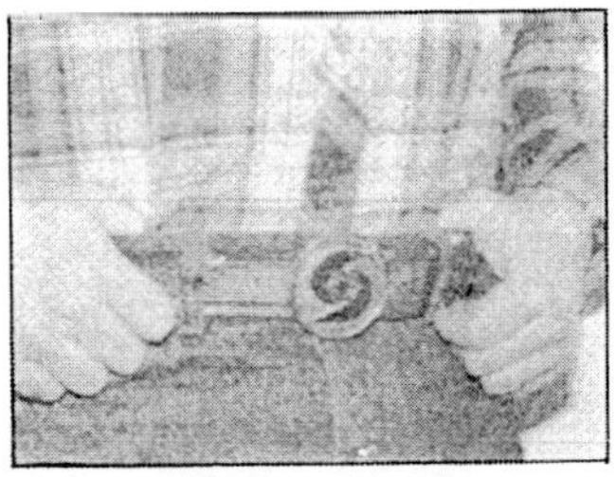

The Cobray Logo Belt Buckle

Cast of solid heavy brass. Fits most of the wide belts available today.

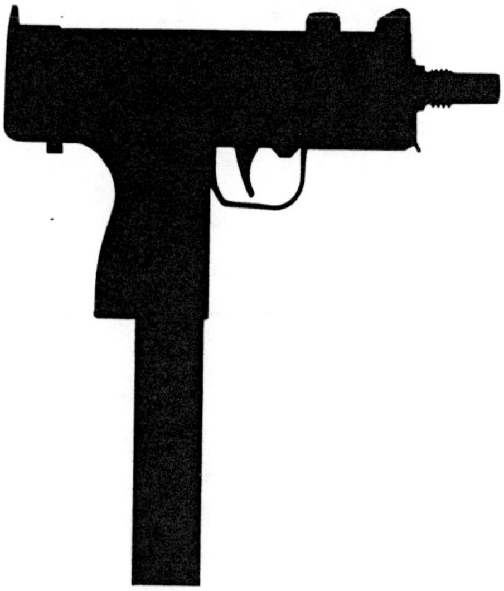

Chapter III
LEATHERWOOD'S MAC
by Randal Stepan

From the outside the only apparent difference is the safety.

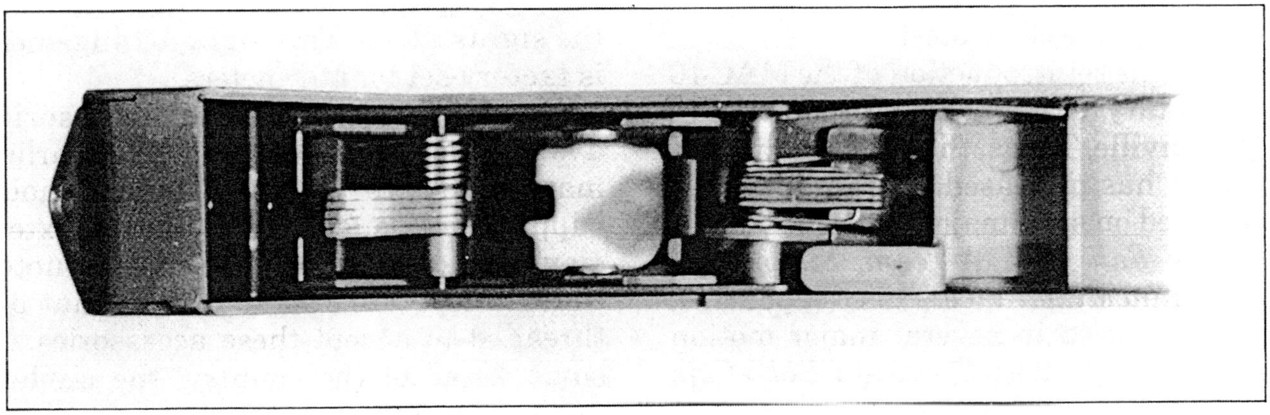

On the inside—a different story! Strange parts in strange places. Note the use of welded laminations in place of machined or investment cast parts.

In 1982, James M. Leatherwood (who in 1971 was the vice president for engineering at Military Armaments Corporation) developed a new machine gun, which worked and resembled the Ingram in appearance. At this time, the tooling for the original Military Armaments Corporation's MAC-M10 Ingram came up for sale. Advanced Armament, Inc. (Leatherwood's company) purchased this tooling.

With the advice of Gordon Ingram and Mitchell WerBell III, James Leatherwood soon had a very efficient working model of the M10A1 for testing, which was superior to the M10's, due to advanced technology and design changes. The weapon worked so well, the decision was made to produce the gun for sale. A new Military Armaments Corporation was set up in Stephenville, Texas. This second generation weapon is available in both .45 ACP and 9mm Luger.

A semiauto version became available to the civilian market in 1983/84, designated: MAC 10A1S, which fires from a closed bolt. Conversion kits available for the weapon can change the MAC 10A1S from .45 ACP to 9mm, by replacing three parts. All versions of the Ingram are made of stamped and welded steel.

Since the reintroduction of the MAC-10 by Military Armaments Corporation of Stephenville, Texas, the popularity of the weapon has increased. The MAC-10 has been used on such major television shows as, *The Fall Guy, A-Team, Simon and Simon* and *Miami Vice*. This weapon has, also, appeared in several major motion pictures: *Death Wish Three* and *Robo-Cop*.

All models of MAC-10 have the bolt handle on top of the receiver. On some models, the bolt handle doubles as a safety. When placed in the "on" position the sight picture is obstructed and the weapon operator can tell at a glance whether or not the safety is engaged. On the semiauto MAC-10, the safety is located in the trigger guard. The model SM10 has the safety located in (and to the side of) the trigger guard, where it is operated by the trigger finger. The safety on the MAC10, model M10A1S is of the Garand type, located in the trigger guard. This safety is, also, operated by the trigger finger.

The magazine well is welded to the lower receiver, making a strong pistol grip. The magazine release is located at the rear bottom of the pistol grip. Due to the weapon's design being so short, proper handling of the weapon is essential at all times. Many shooters elect to use a dummy silencer or barrel extension to create a handhold, which generally improves weapon accuracy when fired from the hip.

The sights on all models of the MAC-10 are the same. They consist of an aperture hole drilled into an extension of the rear receiver plate. The front sight is of the post type and is protected by two ears, one on either side of the front post, which can be adjusted to a limited degree by bending the front post, as needed, or by filing it. Though, in most cases, it is best to leave the sights alone. This sight arrangement is factory-set for 100 meters.

There are a number of accessories available for the MAC-10. As stated earlier, many shooters attach a dummy noise suppressor (a silencer), or a barrel extension to their MAC's. It should be noted that all MAC-10's have barrels that are threaded to accept these accessories. In some areas of the country, the civilian market is allowed to legally-own a functional suppressor, and those who choose to use the MAC-10 suppressor have one of the best available. This is due to the subsonic nature of the .45 ACP round. The success of this suppressor has done much to increase the sales of the Ingram MAC-10.

Other accessories for the MAC-10 are: a series of magazines of varying capacities and a shoulder sling for carrying the weapon, which leaves the hands free. There is, also, a short hand strap that attaches to the hanger behind the barrel, which aids in steadying the weapon when firing. Another accessory that is popular and attractive is the walnut pistol grip, which adds a personal touch that is very appealing to the eye. There are a number of brass catchers on the market that will fit the MAC-10. Several companies in the past have made briefcase type carriers for the MAC and many of its accessories. Currently, soft-sided SWAT type carrying cases are available through Military Armaments Corporation and several aftermarket companies. The F.B.I. rig is, also, available for the MAC-10. This shoulder strap type rig allows the weapon to be carried under an overcoat or top coat. This rig, also, allows for the carrying of two 30-round or 40-round magazines.

It should be noted that a shoulder stock is available as an accessory. These shoulder stocks have been manufactured by several aftermarket firms. These stocks are legal on Title II weapons, only. It is a federal crime to place a fixed or folding stock on a pistol. It should be remembered that, due to their barrel length, semiauto MAC-10's are classified as pistols.

All MAC-10's, regardless of their place of manufacture, are field-stripped in the same manner. This weapon has been designed so no special tools are necessary to disassemble or reassemble it.

First, remove the magazine, cycle the weapon and make sure it is unloaded, place the weapon on safe. If the weapon is equipped with a suppressor or barrel extension, remove it at this time, then the frame pin. (There are several different types of frame pins.) Depending on which type of pin, you may have to remove a screw head, punch the receiver pin catch back or you may have a spring-loaded type. In any case, push the frame pin out, it should be removed from the right side. (In some cases a slight rearward pressure may have to be applied to the upper receiver to allow the frame pin to be easily-removed.) Now, you can lift the upper receiver assembly upward and forward, separating it from the lower assembly. Place the lower assembly aside for cleaning or service.

The upper assembly contains the barrel and the bolt assembly. To remove the bolt assembly, move it to the rear of the receiver, then align the bolt handle with the large opening at the rear of its track. Pull upwards on the bolt handle and remove it. In case of a tight fit, use a punch to remove the bolt handle by tapping it from the bottom. Now, remove the bolt assembly by sliding it all the way to the rear and out of the upper receiver into which the barrel is threaded and not normally removed. The bolt spring, spring guide and ejector are now removed from the bolt. The bolt spring and spring guide are held together by a split ring or "E" clip and usually are not separated. The weapon is now ready for normal cleaning or service and no further disassembly is required. To reassemble the weapon, reverse the disassembly procedure, as outlined above.

Due to the design of the MAC-10, it is a very reliable weapon and will give many years of service, if properly-cared for. As with all weapons, dirt is the biggest enemy. To prevent malfunctions, periodic inspections and proper cleaning is necessary.

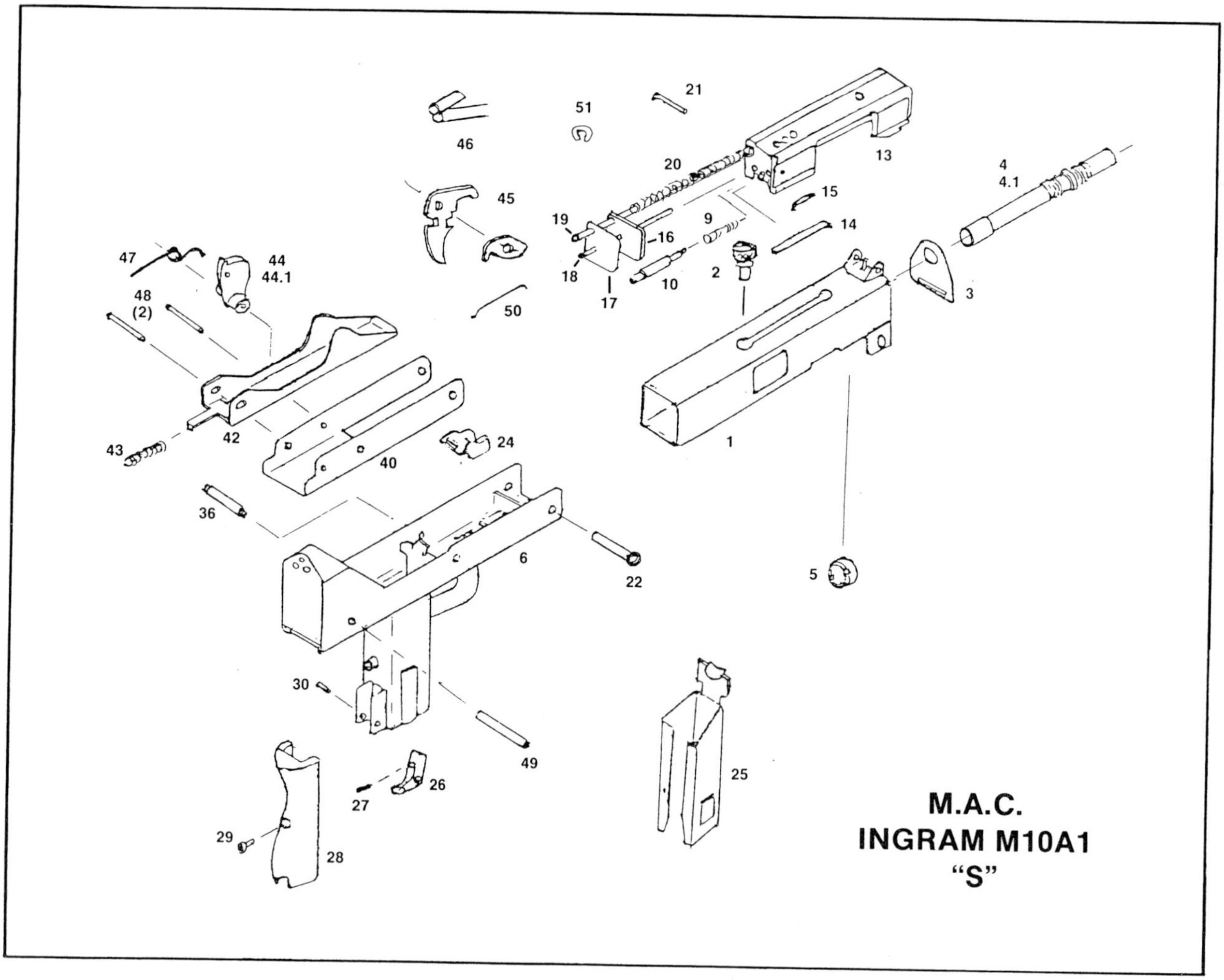

1 Receiver w/Barrel
2 Bolt Handle
3 Hanger
4 Barrel - 9mm
4.1 Barrel - .45 ACP
5 Barrel Nut
6 Frame Assembly
9 Firing Pin Spring
10 Firing Pin
13 Bolt Assembly
14 Extractor
15 Extractor Spring
16 Buffer
17 Plate, Rear Recoil
18 Ejector
19 Guide Rod, Recoil
20 Recoil Spring
21 Extractor Retainer
22 Receiver Pin
24 Safety
25 Magazine Insert & Feed Ramp (9mm only)
26 Magazine Catch
27 Magazine Catch Spring
28 Grip (plastic)
29 Grip Screw
30 Magazine Catch Pin
36 Trigger Pin
40 Lockwork Housing
42 Trigger Bar
43 Trigger Bar Spring
44 Hammer - 9mm
44.1 Hammer - .45 ACP
45 Trigger
46 Trigger Spring
47 Hammer Spring
48 Trigger Bar & Hammer Pin (2 req'd)
49 Roll Pin
50 Retainer
51 Guide Rod Retainer (E ring)

The most common malfunctions associated with the MAC-10 are failure to feed, which is usually caused by a dirty or defective magazine and failure to fire, which is normally-caused by a dirty bolt face (open bolt weapons), dirty firing pin hole (closed bolt weapons) and defective ammunition.

All MAC-10's were designed to fire jacketed round nose bullets. The use of semiwadcutter or hollow point bullets may cause failure to feed problems. Normally throating and polishing of the chamber and feed ramp will solve this feeding problem.

Dirt, grit, grime, unburnt powder and brass particles are all items that are normally found in a weapon after it has been fired. When mixed with the normal oils used to assure smooth operation, a sticky sludge is formed. Besides being unsightly and interfering with normal functioning of the weapon, this sludge causes wear on internal parts, which can, in time, cause a mechanical malfunction of the weapon. The following chart lists some of these and the corrective-actions needed to return the weapon to an operational condition.

As a result of the 1986 machine gun ban, no new select-fire weapons can be sold or registered to the civilian population of the United States. It is still legal to own a select-fire weapon, if it was registered prior to April 1986 and there are no state laws restricting ownership. As a result of this legislation, the Ingram MAC10A1S is the only model of the Ingram available to the general public. For this reason and for purposes of academic information, the procedure for the total disassembly of the MAC10A1S will now be covered.

The total disassembly of any weapon for purposes of repair, maintenance (or any other reason) should be done only by a qualified gunsmith. Any attempt at disassembly of the MAC10A1S by other than Military Armaments Corporation personnel, or one of their authorized warranty and repair stations, will void any factory warranty.

The MAC10A1S should first be checked to make sure that it is unloaded. If a barrel extension or fake suppressor is attached, it should be removed at this time. The weapon can now be field-stripped into its three major components as outlined earlier.

Of the three major components, the upper receiver, which contains the barrel, is the simplest with which to deal. As stated earlier, the barrel is threaded into the upper receiver. It should be noted that the barrel nut is located inside the upper receiver. This is a special one-way nut, which is installed by hydraulic ram. If the upper receiver or barrel is damaged, it must be returned to the factory or a warranty station for repair or replacement.

The bolt assembly is the next to be covered. As noted earlier, the recoil spring guide rod, recoil spring and ejector have already been removed from the bolt during field-stripping. The ejector can be removed by pushing it toward the rear of the recoil plate. Next, compress the recoil spring and remove the "E" clip. Slowly, release the spring pressure. Now the recoil buffer and recoil guide rod can be separated.

The extractor retainer is removed next, by using a punch of the proper size. The extractor retainer is removed from the left side of the bolt. Care should be taken at this point not to lose the firing pin. On some of the older M10A1S's the firing pin is under spring pressure. On newer M10A1S's, the shape of the firing pin has been changed and the firing pin compression spring has been deleted. In either case, the firing pin is removed from the rear of the bolt at this time. By using a screwdriver or similar tool, *carefully* pry forward on the extractor and remove it from the bolt. Care should be taken not to lose the extractor tension spring.

There are two styles of extractors: The old and new style. If you have the old style the extractor end on the outboard side is flat. On the new style extractor, this end is half round looking on its outboard side. As a point of information, the old style extractor is no longer available. A functional extractor for the holder weapons can be made by modifying the new style. This involves removing the excess material from the extractor end on the outboard side, until it is flat. With one exception, the bolt assembly is reassembled by reversing the above procedure. (NOTE: When reinstalling the firing pin, it is important that the one-half round area at the back of the firing pin is facing down, toward the bottom of the bolt. This will allow the extractor retainer to be reinstalled.)

The last section to be disassembled is the lower frame (receiver) assembly. Examine the lower frame assembly and note the relationship of all the parts. Press down on the retainer where it engages the trigger pin with the left-hand, index finger. Using the right-hand, rotate the retainer 90 degrees at the forward end. Using the same hand, pull the retainer forward approximately one-half inch, until it clears the trigger pin. At this point the retainer can be removed by turning it another 90 degrees in the same direction and fishing it out of the retaining hole.

Secure the back of the weapon against your chest or other firm object. Use a nonmarring tool or a blunt screwdriver. Place the tool on the left side of the lockwork assembly, in front of the trigger pin. Push back against the lockwork assembly to compress the trigger bar spring. While keeping the trigger bar spring compressed, use a punch of the proper size and push out the trigger pin to the left and remove it from the left side of the frame. While holding your thumb over the trigger, remove the punch out from the right side. With a nonmarring tool or screwdriver push back on the lockwork housing and raise it halfway out of the receiver. Now, remove the trigger and trigger spring through the top of the frame. After the trigger and trigger spring have been removed, push back on the lockwork housing (to compress the trigger bar spring) and raise the front of the housing until it will clear the frame. Remove the lockwork housing and lay it aside. The Garand type safety can now be removed by bringing it all the way to the rear of its slot and lifting upward and out.

Use an Allen wrench of the proper size to remove the grip screw. Be careful not to lose the magazine catch spring when removing the grip. Using a small punch, push out the magazine catch pin and remove the magazine catch. The frame is now completely-disassembled with the exception of the frame roll pin.

To disassemble the lockwork housing, remove the trigger bar spring. Using a punch of the proper size, push the rearmost trigger bar/hammer pin out to the left. Using the same punch, push the remaining trigger bar/hammer pin out of the lockwork housing. Now, the hammer and the hammer spring can be removed as a unit. The trigger bar and the lockwork housing can be separated at this time.

When reassembling the frame, the Garand type safety switch is installed first. Then the reassembled lockwork housing assembly is placed back into the frame, until it is flush with the top of the frame sides. The trigger and trigger spring are now installed as a unit, making sure that the ends of the trigger springs are positioned on top of the safety switch. After this trigger spring and trigger are seated in the weapon, push the lockwork assembly the rest of the way in, so that it fits flush with the bottom of the frame.

Using a punch of the proper size, insert it in the trigger pin hole from the right side, while aligning the holes in the right side of the lockwork, the safety bushing and safety rocker, right side of the trigger spring and the trigger. Continue pushing the punch through to pick up the left side of the trigger spring, lockwork housing and trigger pin hole in the left side of the frame. Now, push the trigger pin in from the left side of the frame, while withdrawing the punch from the right side of the frame. At this point, the retainer is installed in its proper location. The inboard end of the retainer is depressed into its slot in the trigger pin. Install the magazine catch in its location in the handle and reinstall the magazine catch pin. With the magazine catch spring in place, install the pistol grip and tighten the pistol grip screw to hold the pistol grip to the weapon.

Follow the reassembly for field-stripping to place the weapon back in operating condition.

The following companies have been of help in preparing this information: Military Armaments Corp. of Texas, Defence System International of Georgia, Stepan's Auto Service of California and the Bureau of Alcohol, Tobacco and Firearms (Los Angeles office).

AUTHOR'S POSTSCRIPT

During the time this author spent researching this book, a number of attempts were made to contact the designer of the Ingram M-10 weapon system. Mr. Ingram and I finally met at one of the major gun shows held in Southern California. As a result of this meeting this author has had the occasion to interview Mr. Ingram personally. Mr. Ingram states, "There is no relationship with the MAC-Leatherwood-WerBell group of Stephenville, Texas." Mr. Ingram informed this author that, "He is displeased with certain design changes on weapons bearing his name."

After his departure from the original Military Armament Corporation, Mr. Ingram has been working on several projects for companies overseas. Since his return to this country he has been developing a new, greatly improved, line of submachine guns, carbines and pistols. These new weapons are chambered in .45 ACP, 9mm Parabellum, .22 Long Rifle, as well as several unqiue new magnum cartridges of his own design. This author will present full details on these and other developments relating to the new "Mark Series" of weapons from Ingram Engineering Services as they become available.

	Powder Springs MAC-10 Open Bolt	Stephenville MAC-10 Open Bolt	Stephenville MAC-10 Closed Bolt
Barrel Length:	5.75" Both .45 & 9mm	Same	Same
Length (Stock Folded):	10.5" Both .45 & 9mm	Same	Same
Length (Stock Extended):	21.5" Both .45 & 9mm	Same	N.R.
Magazine:	30 Rds. .45 ACP 32 Rds. 9mm	Same	Same
Muzzle Velocity:	948 fps .45 1280 fps 9mm	Same	Same
Cyclic Rate:	.45 = 1036 rpm 9mm = 1123 rpm	.45 = 934 rpm 9mm = 759 rpm	.45 = 186 rpm 9mm = 186 rpm Rapid Fire
Weight Unloaded:	6.25 lbs.	Same	Same

MALFUNCTION	POSSIBLE CAUSE	CORRECTIVE ACTION
Failure to Feed:	Weak or bad magazine Worn magazine notch Bad ammunition Worn or broken mag. catch	Replace magazine Replace magazine Replace ammunition Replace magazine catch
Failure to Chamber:	Dirty chamber Blocked chamber Weak recoil spring	Clean chamber Remove blockage Replace recoil spring
Failure to Fire:	Bad ammunition Damage/broken firing pin Weak recoil spring Worn hammer bar Worn/damage lockwork assy.	Replace ammunition Replace bolt or firing pin Replace recoil spring Replace lockwork assy. Replace lockwork assy.
Failure to Extract:	Broken extractor Weak extractor Broken extractor spring	Replace extractor Replace extractor Replace extractor spring
Failure to Eject:	Broken ejector Damage extractor	Replace ejector See above
Failure to Cock:	Worn sear Worn sear notch Bent guide rods Worn lockwork housing assy. Low-powered ammunition	Replace sear Replace bolt Repair/replace guide rods Replace lockwork housing assy. Replace ammunition

Again, periodic inspection and proper care will reduce the possibility of malfunctions.

Chapter IV
JERSEY ARMS
by Gary Reisewitz

Just when it seemed that all of the new MAC-10's that were ever going to be available at reasonable prices were gone forever, something new has shown up on the market that brings back fond memories of the recent past. This is the Jersey Arms' .45 caliber submachine gun that bears an absolutely-striking resemblance to the RPB type MAC-10 submachine guns. These are being sold by Cobray, Inc. of Smyrna, Georgia, at a very reasonable dealer price. Having seen an example of the weapon, I began to notice a few peculiarities that further piqued my curiosity. I, then, decided to find out where these weapons had been for the past seven or eight years and the extent that they differed from the Powder Springs' or RPB M-10. The history and source of these "new" submachine guns proved to be quite interesting.

Prior to 1982, RPB, Inc. was manufacturing the open bolt semiautomatic version of the M-10 and M-11 as pistols and carbines. They were not the only people who thought that this was a good idea. A company known as Hatton Industries thought it was a good enough idea to give it a try themselves. Under the name of Jersey Arms, in early 1982, they pressed out the frames of what were to be open bolt semiautomatic pistols that would somewhat resemble the M-10. However, the BATF did not think that this was such a good idea and in 1982 informed both RPB, Inc. and Hatton Industries that they could no longer do this. None of the Jersey Arms' frames were ever completed into pistols and the BATF permitted all of these frames to be classified as machine guns. This was no longer the simple profitable venture that the Jersey Arms folks thought it would be, and they decided not to continue. The total production of the Jersey Arms' frames was 520. They lay dormant for a number of years until S.W. Daniels, Inc. negotiated the purchase of the frames from Hatton Industries.

When S.W. Daniels purchased the frames, a number of modifications had to be made in order for the weapons to more-closely resemble the RPB, Inc. version of the M-10. The Jersey Arms' frames had a different type of rear sight welded to the frame. This had to be removed and the original M-10 type sight plate welded in place. The rear stock latch pin holes had to be welded up and redrilled to permit the stock latch mechanism to be welded in place.

The sear pin hole had to be drilled and the M-10 type trigger guard had to be installed. The weapons were completed with both, old original RPB M-10 parts, and some of recent manufacture. There were a total of 420 submachine guns assembled, all of which were in .45 ACP caliber. It should, also, be noted that S.W. Daniels, Inc. is managed by the same folks who ran RPB, Inc. when it was manufacturing the RPB M-10 submachine guns.

There are still some differences between the Jersey Arms' weapon and the RPB, Inc. M-10. The most obvious is the type of buttstock that is used, which is noticeably different from the old standard wireform stock, a design carried over from the Powder Springs' guns. When extended, it does lock in the open position, but when closed, it stays closed only due to friction. Therefore, the stock latch on the underside of the receiver does not have to be pressed in order to extend the stock. When the stock is extended, it is approximately 2.75 inches shorter than the original type wireform stock. When collapsed, the weapon is approximately 1.5 inches longer than if equipped with the older style stock. The greatest advantages to the new stock design are the increase in rigidity and speed with which it can be extended and made ready for use.

If desired, however, the original wireform stock can be placed on the weapon by merely removing one stock and inserting the other.

Another difference lies in the magazine catch. The catch and well are designed to accept the unmodified M-3 Grease Gun magazine. However, most M-3 magazines still require a light beveling of the rear magazine catch edge in order for the catch to secure them. It is a simple, two-minute job to remove the magazine catch, altogether, and replace it with an original M-10 type catch, which will enable the weapon to share magazines with the rest of the M-10's of the world.

The remaining significant difference is that the takedown pin that secures the upper receiver is of a two-piece design, very similar to the one used on the new Cobray M-11/9 pistol and submachine gun.

Other than these differences, the weapon is almost a carbon copy of the RPB, Inc. M-10 submachine gun. The weapon fieldstrips the same and, other than the markings, would be very difficult to differentiate from one of the earlier weapons. This is not too surprising since the guns were essentially-made by the same people.

The Jersey Arms' submachine gun is as much a "MAC-10" as any other variation, and possibly more than some. The only serious complaint that I can voice is that the appearance of the welds and general fit of the parts are a little lower in quality than that of the Powder Springs' guns, although it is on par with many of the RPB specimens that I have seen.

It is interesting to note that with a total production of only 420 weapons, the Jersey Arms' gun may be the lowest production quantity of any of the clearly identifiable variants and, in the future, may be worth more than the other, generally more expensive ones.

RANGE TEST

Aside from researching the various differences between this and other types of MAC-10 submachine guns, I, also, wanted to know how well the Jersey Arms' gun fired. I went to the local range with a few hundred rounds of .45 ACP reloads, several 30-round and three 40-round magazines. The reloads were assembled with 230-grain round nose, cast lead bullets loaded to resemble the standard military hardball load. Some may question my use of cast lead bullets in a weapons' test, but it has been my experience that the real *shooters* out there use the cheapest fodder that they can find and shoot lots of it. I decided that the average, real shooter is going to be using these cheap cast lead slugs, so that is what I should use to test the weapon. It, also, stands to reason that any machine gun that will shoot cast bullets will shoot the jacketed ammo just fine, but that is not always true.

Arriving at the range, I began my function tests with some silhouettes I had set up at the 25-meter line. From prior experiences with these types of weapons, I knew that any attempts at accuracy testing were going to be rather pointless. I decided to just see how well the weapon fired.

After consuming all the ammunition that I had brought, I had yet to experience any malfunctions with the 30-round magazines. However, I was never able to get any of the 40-round magazines to reliably function. Upon further examination and by attempting to manually strip rounds from the 40-round magazines, I came to the conclusion that the fault was with the magazines (and not with the weapon). None would reliably permit even the manual-stripping of rounds from the lips, without the remaining cartridges binding within the magazine and failing to rise to the feed position. This may have been due to the greater friction caused by using the cast lead bullets. These types of magazines might work well with full metal jacket ammo. The 30-round magazines were a collection of modified military grease gun magazines and aftermarket MAC-10 magazines and all performed perfectly.

The rate of fire was in the 900+ rpm range, and the weapon was controllable enough to be safe. It wasn't a Thompson or a Sten, but it was far more controllable than a machine pistol. I did manage to hit a few of the silhouettes with a few rounds, but not consistently enough to say that I would trust it at that range.

The MAC-10 has always been and will continue to be strictly a close range weapon. I don't recall ever having any reliability problems with any variants of the MAC-10. Some may have been more aesthetically-pleasing than others, but they all shot without a hitch. The Jersey Arms' submachine gun certainly lives up to the performance standards of its predecessors.

SPECIFICATIONS

Length, stock collapsed:	13 inches
Length, stock extended:	19.25 inches
Maximum width (takedown pin):	2.25 inches
Barrel length:	5.75 inches
Weight, empty:	6.25 pounds
Type of fire:	Semiautomatic or fully-automatic
Cyclic rate of fire:	Approx. 950 rpm
Magazine capacity:	30 rounds

(Aftermarket magazines in 40-round capacity are readily available, though not always reliable.)

Detail of receiver markings—right side.

Detail of receiver markings—left side.

Chapter V
PARTISAN AVENGER
— PISTOL —
by Nolan Wilson

Again, the same "MAC" appearance with the exception of improved front and rear sights (who uses sights on a MAC?) and the pull ring on the takedown pin.

TOP – PARTISAN AVENGER with
barrel extender
BOTTOM – PARTISAN AVENGER
with clip grip
(360° rotation)

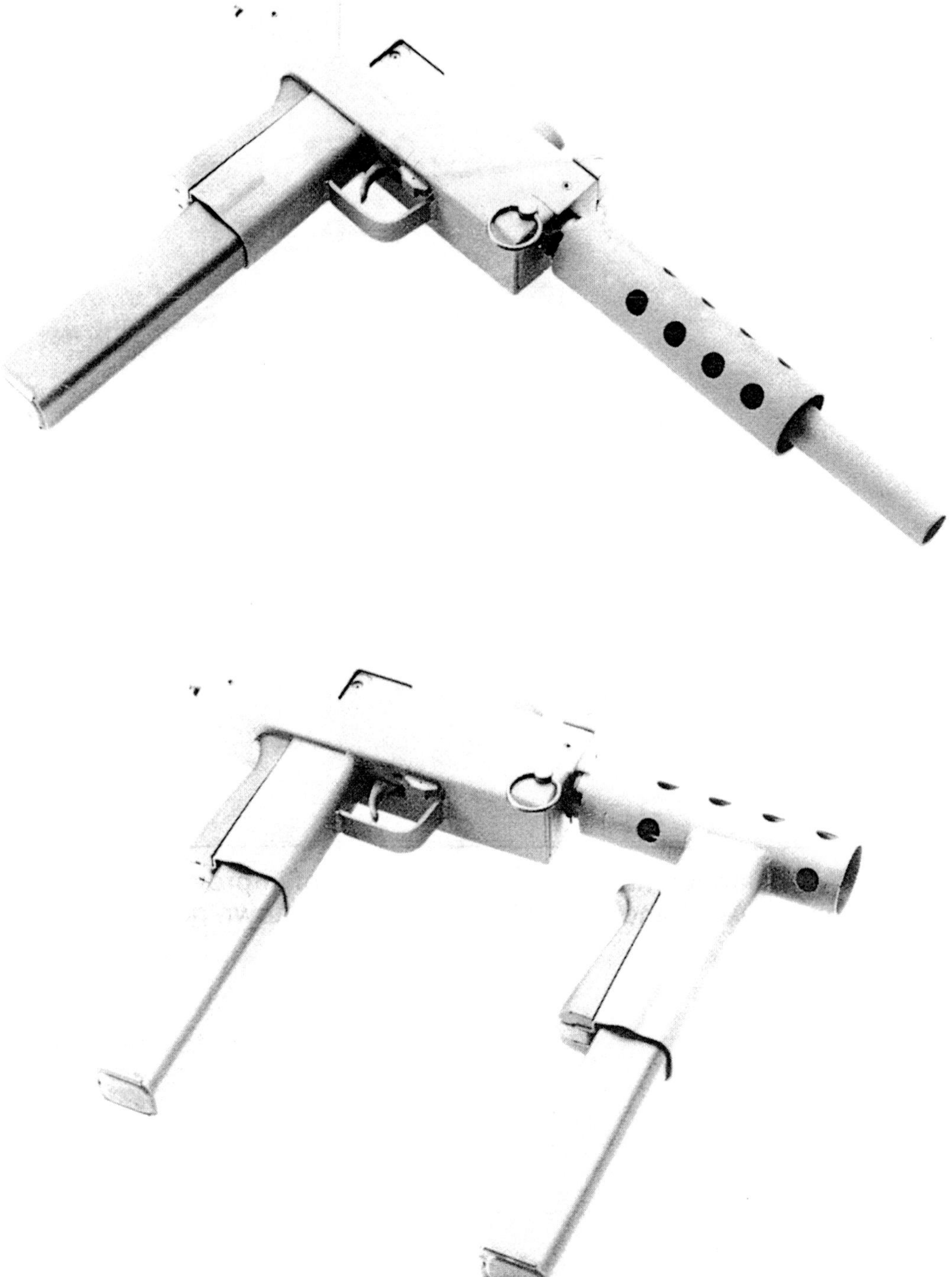

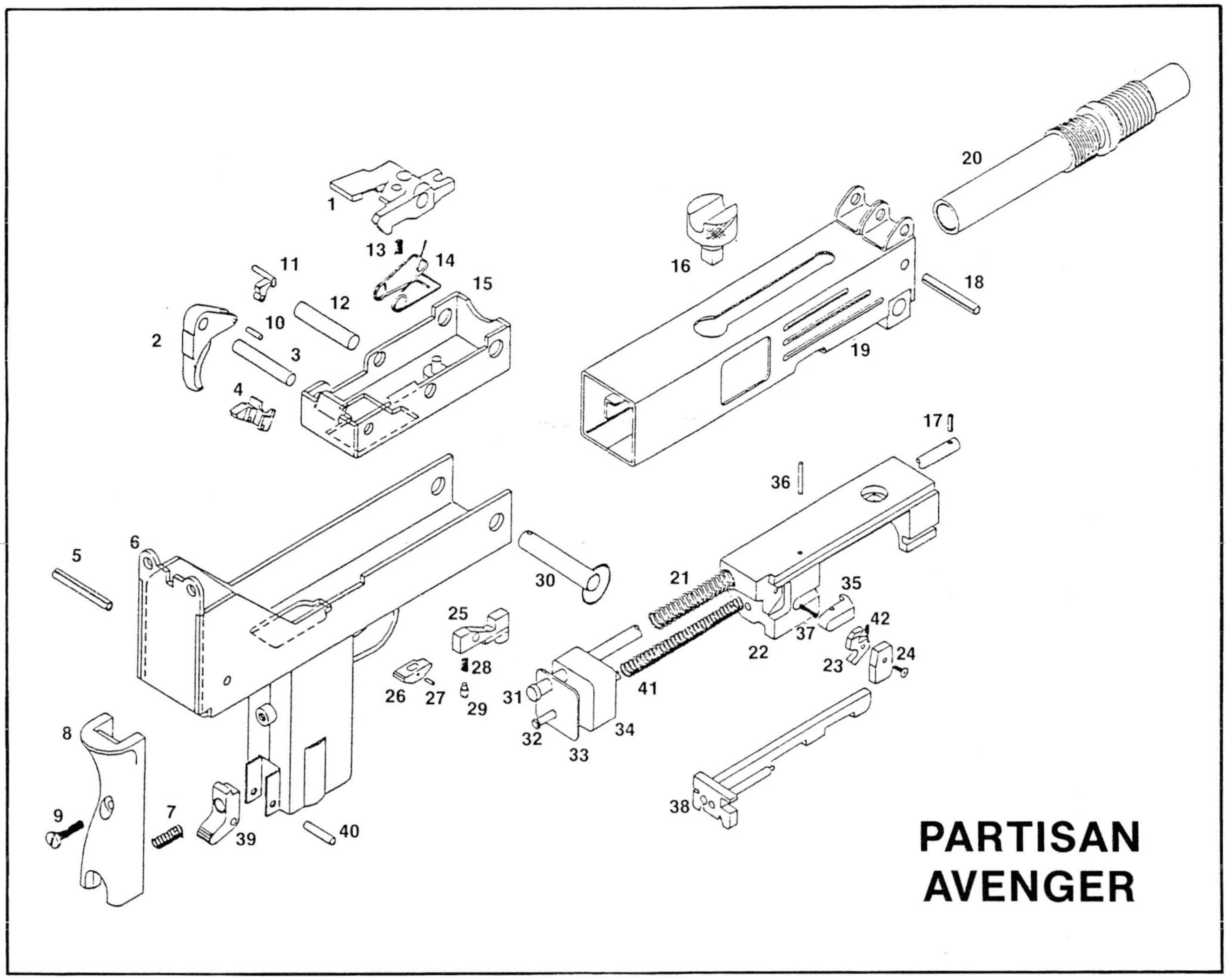

1 Sear
2 Trigger
3 Trigger Pin
4 Sear Cam
5 Buffer Guide Pin
6 Frame Assembly
7 Magazine Catch Spring
8 Plastic Grip
9 Grip Screw
10 Disconnector Pin
11 Disconnector
12 Sear Pin
13 Sear Spring
14 Trigger Spring
15 Assembly Module
16 Bolt Handle
17 Guide Pin Lock
18 Head Spacing Pin
19 Receiver Frame Assembly
20 Barrel
21 Recoil Spring
22 Bolt
23 Dog
24 Retainer Plate & Fastener
25 Safety Slide
26 Safety Button
27 Safety Pin
28 Safety Spring
29 Detent Plunger
30 Front Frame Assembly Pin
31 Guide Rod
32 Ejector Rod
33 Buffer Support Plate
34 Plastic Buffer
35 Extractor
36 Extractor Pin
37 Extractor Spring
38 Firing Pin (striker)
39 Magazine Catch
40 Magazine Catch Pin
41 Ejector Spring
42 Dog Spring

Notes

Notes

Notes

Notes